E. Small

Anatomy and Physiology Rendered Attractive

And the Laws of Health Made Plain, in Conversations between a Physician and his Children.

E. Small

Anatomy and Physiology Rendered Attractive
And the Laws of Health Made Plain, in Conversations between a Physician and his Children.

ISBN/EAN: 9783337232139

Printed in Europe, USA, Canada, Australia, Japan

Cover: Foto ©berggeist007 / pixelio.de

More available books at **www.hansebooks.com**

ANATOMY AND PHYSIOLOGY

RENDERED ATTRACTIVE,

AND

THE LAWS OF HEALTH MADE PLAIN,

IN CONVERSATIONS

BETWEEN A PHYSICIAN AND HIS CHILDREN.

DESIGNED FOR SCHOOLS AND FAMILIES,

AND FOR GENERAL READING.

BY

E. SMALL, M. D.,

AUTHOR OF "A TREATISE OF INFLAMMATORY DISEASES," ETC.

BOSTON:
DEGEN AND ESTES,
No. 23 CORNHILL.
1864.

Entered, according to Act of Congress, in the year 1868, by
E. SMALL,
In the Clerk's Office of the District Court of the District of Massachusetts.

DAKIN, DAVIES, & METCALF,
Stereotypers and Printers,
37 CORNHILL, BOSTON.

PREFACE.

The origin of this work may be briefly stated. Within the last ten years the author has been requested, at various times, to address assemblies, particularly Sabbath Schools and Lyceums, upon the subject treated in this book.

The thought and study necessary to the preparation of these lectures, have impressed him with the belief that the numerous works on Anatomy and Physiology, designed for the use of schools and for popular reading, have been essentially deficient, inasmuch as reference is seldom made in them, to the proofs that stand out so prominently in every part and organ of the human body, of the wisdom and goodness of God in its formation.

This, perhaps, should not be expected in purely scientific works upon these subjects, designed solely for the Student of Medicine; yet, even in these, an occasional reference to the abundant proofs of creative goodness, would not be out of place.

But in works designed for popular instruction, such deficiency is certainly a defect.

The inhabitants of the forest—the aborigines of this continent, may pass unnoticed, or gaze with stupid indifference, upon the stately specimens of artistic skill which adorn large cities; and in them it is excusable. But educated and Christian communities, with works of science in one hand, and in the other the book of inspiration, which asserts that "our bodies are temples of the Holy Ghost," are certainly without excuse, if, while they admire the works of art, they fail to observe in this *living* "temple" the exquisite workmanship of the divine Architect, and to trace out the benevolent designs in its structure, the proofs of which are so abundant. Yet, if authors fail to bring these impressive proofs of divine goodness prominently before the public eye, it cannot be expected that they will be discovered and appreciated by the young, or by the popular mass.

But no such work has ever fallen under the observation, or come to the knowledge of the author; and being impressed with the belief that benefit would accrue to the public, especially to the young, from such a work, in a moral as well as literary sense, has prompted the effort, the result of which is here presented to the public, whether for its approval or condemnation, will be subsequently seen.

The author has deemed it proper to present what to him appears incontestable proofs of the existence, and wonderful capacities, of the human soul. He has also attempted to show, what, perhaps, none will deny, and yet what few fully appreciate, that all things, animate and inanimate, mind and matter, are under the control of established law; that health and vigor of body and mind, depend upon strict obedience to laws which control animate nature; and that all suffering is the result of the violation of law.

The laws of health have been carefully considered, what constitutes their violation pointed out, and directions given for their observance.

After due deliberation the author decided to adopt the colloquial or conversational style, not, however, without being aware of the objections to, as well as the advantages of, such a style.

The author has endeavored so to present the subject as to combine pleasure with the acquisition of knowledge, and render study and reading at once agreeable and instructive.

In most works of this character, technical terms constitute a serious obstacle to the common reader. In this, their use has been avoided as far as is deemed profitable. Anatomical words have been divided into syllables, and those accented, marked.

While this work is not offered as a *complete* treatise upon the subject, it is believed to embrace all the leading and prominent facts, leaving out only the minute and abstruse, which could be of no benefit to the popular reader or young student, and could interest only the professional man.

The author is fully aware that this little work is open to criticism. That it is free from faults he cannot indulge a hope; and all he asks is, that the public will exercise candor, and award to him the credit of an honest intention, and a desire to contribute his mite to the general good, and especially to that of the young.

If it shall serve in any degree to excite thought, and give additional interest in the study of Man, physically, mentally and morally, and impress upon the young the conviction of God's goodness in their formation, and awaken stronger desires for greater knowledge upon this subject, the labor bestowed will not have been in vain.

It will be seen that this work is fully illustrated with beautiful engravings. These illustrations are taken from a valuable work recently published, entitled "The Household Physician," by Dr. Ira Warren of this city, to whom the author's acknowledgements are due.

Boston, Mass., Sept., 1863.

TO TEACHERS AND PARENTS.

The author believes he states a fact from which few will dissent, when he asserts that Anatomy, Physiology, and Hygiene, in the manner in which they have been presented to the public, particularly to the young, have been dry and uninteresting subjects, and to many, absolutely repulsive. But the author has long been satisfied, that it is not the *nature* of the subject, but the mode of presenting it, which has produced this impression upon the mind. Few subjects can be proposed for the study of the young, or persons of any age, which contain so many facts of thrilling interest, when properly presented, as the sciences I have named.

The *importance* of correct and familiar knowledge relative to the structure of our physical system, the various functions performed by its numerous organs, the laws which control these functions, and the means best adapted to preserve them in a healthy condition, every thoughtful person will admit. Deeply to be regretted, then, is the fact, that a subject of such vast moment to man's best interests, and capable, when rightly presented, of affording pleasure and delight, as well as valuable instruction, should be presented in such a manner as to fail to secure these important ends.

The present work, to which your attention is most respectfully called, is an attempt,—how successful others must judge—to remedy the evils here referred to.

The colloquial style, the author is aware, is a novel one in which to present such a subject, but he ventures to hope that it will be interesting to the young, and that the book will be found in some good degree, to fill the desideratum which has so long existed, and has been so deeply felt.

An important advantage in the colloquial style is, that to answer the interrogations, renders it proper and necessary to explain in the most familiar manner, and to descend to minute particulars to an extent that could not, with propriety be done in any other style of writing; thus not only allowing, but rendering it necessary to use familiar language and the most natural mode of expression in explaining the numerous topics dwelt upon.

The division of technical words into syallables, with the accented ones marked, as has been done in this work, will enable even the ordinary reader to give a correct pronunciation, to acquire which is of great importance.

On account of the familiar style, the author has not deemed it necessary to append questions. If, however, it should be found on trial that questions would aid the teacher and parent and facilitate the progress of the pupil, they will be added in future Editions.

TESTIMONIALS.

A TREATISE ON PHYSIOLOGY, BY E. SMALL, M. D.—I have read this work with great interest, and think it admirably adapted to the use of schools and families—particularly, with perhaps some abridgement, to the Sunday School Library. It presents a correct system of Anatomy, Physiology and Hygiene, in a style to gain the attention of the young; and in language simple, flowing and easy of comprehension. Every person who reads the work carefully will be struck with the vast number of ways in which it sets forth the Divine Wisdom and Goodness in the structure of the human system. Into the heart of every child, who becomes familiar with its contents, will sink the lesson that he is the workmanship of a Father, who seeks only benevolent ends by wise and benevolent means. IRA WARREN, M. D.

BOSTON, Dec. 7, 1863.

BIGELOW SCHOOL, BOSTON, July 7, 1863.

I have thoroughly tested a manuscript work on Physiology, by E. Small, M. D., in my class of fifty pupils,—the first class of the Bigelow Grammar School, and I regard it as a work for the young, above praise.

It fills a vacuum in that science which every earnest teacher has so often realized. Some of the chapters explore a new field—one which no other author has ventured to enter.

Most happily has the author revealed the divinity of the human system. I know of no work so admirably adapted to teach, not only Physiology, but a pure Theology to the young, as this. I long to see it in print, and again avail myself of its able assistance in teaching that interesting science. This work is full of "*object lessons*" for schools, lessons divested of the stiffness and formalities which too often characterize school books.

Physiology in this work is presented to the mind in an *attractive manner*, the conversational style, which is peculiarly suited to the tastes of children. Some of my pupils who thought Physiology and Hygiene to be dry studies, have learned by the aid of this manuscript, to love them, and look forward with eagerness to the recitations.

TESTIMONIALS.

While I regard this work as admirably prepared for a text book in teaching Physiology in schools, it is equally well adapted to instruct and interest families and the common reader. I know of no work for children that so beautifully portrays the divine character. In every topic the All-Wise Father is so revealed that the heart of the child spontaneously flows forth in devout gratitude and affection to the Author of his being.

R. C. MATHER,
Head Assistant in the Bigelow School.

[From Rev. L. D. BARROWS, D. D.]

Nearly all our text-books on Physiology, being prepared for the medical profession or the high schools, are so scholastic and technical that they are altogether above the comprehension of the common masses of our youth. How then shall they become informed on this subject? It seems an absolute necessity that some work should be prepared for them, divested, as far as possible, of technicalities, and whatever else renders the study of the subject dry. This is the precise demand which the author of this work has most happily supplied. A child with ordinary common school intelligence, at ten or twelve years, will understand every paragraph of it. And being put in the conversational form, renders it still more attractive to the young. Thus it will be found, whether this volume is used as a text-book in the schools, or for family reading, it will not fail to take a deep hold of all inquiring minds.

Something also must be done to clothe science and useful literature with more attraction, or we can never compete with the intoxicating excitement of romance. This book is a step in the right direction. Another impressive fact is,—the author has made it cardinal in this work to keep God before his readers and pupils, both in his wisdom and goodness.

There is also, a loud call for some treatise on Physiology quite different from anything now before the people, which shall be more *practical* in character. It is well to know how many bones we have, and how they are put together; but how much more health and happiness depend on our knowledge of the proper care and use of the stomach; and what relation diet, exercise, clothing, bathing, rest, etc., sustain to health. This, perhaps, is one of the crowning excellences of this book. Full and important details are found in it of practical instruction for prevention of disease, suffering and premature death. These numerous sanitary suggestions are worth more than the cost of the whole work; and well studied in families, will prevent much suffering. No family can afford to be without it. It will prove a pleasing and cheap substitute for medicine.

BOSTON, November, 1863.

CONTENTS.

Conversation		Page
I.	INTRODUCTORY	17
II.	THE BONES AND JOINTS	21
III.	JOINTS OF THE BODY (Continued)	45
IV.	THE MUSCLES	57
V.	THE MUSCLES (Continued)	71
VI.	THE NERVOUS SYSTEM	82
VII.	THE NERVOUS SYSTEM (Continued)	98
VIII.	DIGESTION AND NUTRITION	108
IX.	DIGESTION AND NUTRITION (Continued)	132
X.	THE BLOOD-VESSELS AND LUNGS	140
XI.	THE CIRCULATION OF THE BLOOD	155
XII.	THE CIRCULATION OF THE BLOOD (Continued)	170
XIII.	THE SECRETORY ORGANS OR GLANDS	187
XIV.	ABSORPTION	199
XV.	THE SKIN	214
XVI.	THE TEETH	228
XVII.	THE SPECIAL SENSES—THE SENSE OF VISION	243
XVIII.	SENSE OF HEARING	265
XIX.	THE SENSE OF SMELL	280
	THE SENSE OF TASTE	284
	THE SENSE OF TOUCH	291
XX.	MISCELLANEOUS	296
XXI.	THE SOUL	312
XXII.	LAW	327
XXIII.	HYGIENE—THE PRESERVATION OF HEALTH	346
XXIV.	HYGIENE	361
XXV.	DIET	382
XXVI.	HYGIENE	413

ANATOMY AND PHYSIOLOGY.

CONVERSATION I.

INTRODUCTORY.

THE CREATOR'S POWER DEMONSTRATED BY HIS WORKS—HIS GOODNESS AS CLEARLY PROVED BY HIS WORKS AS HIS POWER.

Father.—My dear children, I am happy to meet you again in our study, where we have often conversed with so much pleasure, and, I trust, with profit.

Frederick.—We are all here, Father, waiting with much interest to hear what you will say on the new subject about which you promised to speak at the next meeting.

Frank.—I feel sure we shall be delighted.

Mary.—I am anxious for you to commence at once.

Father.—You must be patient, my children, if you would obtain profit; to do which, we must be systematic, and proceed in the order due to such a subject.

Fred.—I know, Father, you are always for reducing every thing to system and order. I feel desirous to know what proof you can give of God's goodness in the formation of our bodies. I should be glad to have proof that God is good, for in our last conversation we were alarmed at what you told us of the power of God, so that I am really afraid to pray.

Father.—What did I say, my son, that frightened you?

Fred.—You said that this earth on which we live, which is more than twenty-five thousand miles in circumference, is moving in its orbit round the sun at the rate of more than one thousand miles a minute.

Father.—That is so, my son; and that you may be able to form a better estimate of such amazing velocity, I will here state what I then neglected to say, that the speed of the earth in its annual revolution round the sun, is more than one hundred times greater than that of a ball shot from a cannon, which is about six hundred miles an hour, or ten miles a minute. In other words, while a cannon ball travels one mile, the earth one hundred.

Frank.—Is that possible?

Father.—It is not only possible, but it is a *fact*, which may be easily demonstrated. The earth is not only moving at this prodigious rate, but the moon, which revolves round it at the distance of two hundred and forty thousand miles, is also carried along with it. At the same time the earth revolves on its own axis from west to east, every twenty-four hours; and all

Creator's power and goodness.

these wonderful and complicated movements are so uniform and regular, that not a jar is produced, nor can we realise that the earth moves at all.

Frank.—I am startled, Father, at your statements!

Father.—As startling as this appears, it is still a more thrilling fact, that there are countless millions of worlds, some a thousand times larger than our earth, which are flying through infinite space with a surprising degree of velocity, and with the utmost exactness and order. It is certainly very proper that we should feel the most profound awe when we contemplate the works of God; and could we not discover proofs of goodness as well as power in them, we might well fear and tremble.

Frank.—That is just what I have been thinking, Father, since our last conversation. For, if such power is not controlled by goodness, how do we know that it will not be exerted for our destruction seems to me that it would be cause for terror, . than love.

Mary.—It appears to me just as Frank says. I cannot doubt the power of the Creator—his works prove it. But do his works prove his goodness as clearly as his power?

Father.—I think so, my daughter. And by a careful study of his works, especially in the structure of our bodies, we shall see that his goodness is as fully demonstrated as his power. And what can be more ennobling than such a study? Or what can inspire a greater degree of devout adoration to the God who made us.

At the close of our last interview I promised to commence a series of conversations at our next meeting, in which I was to prove the wisdom and goodness of the Creator by the structure of the human body, which promise I shall now attempt to redeem. And if the display of power impresses us with a feeling of awe, the manifestation of goodness should certainly inspire us with love.

Fred.—So it seems to me. And I have sometimes thought, when reading the book of Exodus, that could I have stood with Moses on Mount Sinai and heard God speak,—or could I have seen the tables of stone on which he wrote the ten Commandments, I should ever after have felt the most profound awe and adoration.

Father.—But, my son, let me assure you that the finger of God has left its impress upon your own body, as clearly traced as were the ten commandments upon the tables of stone; nor should the adoration be less profound in the one case than the other; for it is the finger of God in both; and where God is seen, or leaves his impress, "*let man adore.*"

CONVERSATION II.

THE BONES AND JOINTS.

ANATOMY, PHYSIOLOGY AND HYGEINE DEFINED—NAME AND DEFINITION OF PARTS OF WHICH THE BODY IS COMPOSED—OSTEOLOGY—NUMBER, NAME, AND LOCATION OF THE BONES—USES OF THE BONES—JOINTS OF THE BODY—SINOVIAL FLUID, OR JOINT WATER.

Children.—We are again waiting very impatiently, Father, hoping that this time you will tell us something of what you say is so wonderful in the formation of our bodies.

Father.—I am pleased, my dear children, to find you so much interested in the subject upon which we are to converse, and I hope you will not be disappointed in the pleasure and profit you anticipate. But I must again check your impatience, before I commence to give the description of our bodies that I have promised. System and order are Heaven's first law, in all His works, and it should certainly be ours in attempting to investigate them. I must here state, at the begining, that I do not intend, in these conversations which I have promised you, to go into the *minutiæ* of the Anatomy and Physiology of the human body; yet, to talk intelligibly, we must have some knowledge of these subjects. I am well aware that *Anatomy, Phy-*

siology and *Hygiene*, are usually regarded by young people as dry subjects, and destitute of interest, if not absolutely repulsive. But this is an erroneous view, and is the result of a wrong presentation of the subject to the youthful mind. Of the truth of this statement I hope to convince you, before we close this series of conversations. Of the importance of such knowledge, no thoughtful person can indulge a doubt. In what I intend to say I shall pursue the following course :—I shall first briefly describe the *Anatomy* and *Physiology* of each part and organ of the body upon which we speak, as the bones, muscles, nerves, stomach, heart, lungs, eyes, ears, teeth, skin, lymphatics, arteries, veins, capillaries, &c., after which I shall explain the laws which control the health of the various organs and parts of the body.

Mary.—I did not suppose, Father, that you were going to talk to us about the bones, the heart, the veins, and such things, for it always frightens me to hear about them. I thought you would tell us some wonderful things about the formation of our bodies, and show us that God is good to make us as we are.

Fred.—So I thought.

Father.—So I will; but in order to do so, I must first describe the parts of which our bodies are composed. Should you see a machine in operation which indicated great wisdom and skill in the one who invented and constructed it, the complicated movements of which you felt desirous to understand, but which you could not comprehend without a careful study of all its parts separately, that you might see the connection and influence each part has with, and upon the other, in producing the results you witnessed, what would you do?

Fred.—I should examine and study the parts separately, of course.

Father.—If, then, to understand a complicated machine, made by man, you would examine the parts separately, *how much more necessary* is it, in order to comprehend the infinitely more complicated piece of mechanism God has made—our bodies—to study *its* numerous parts and organs, and their movements, separately?

Fred.—I suppose it must be so, and to this I have no objection. But you said you would describe the *anatomy* and *physiology* of the numerous organs of the body.

Father.—So I did; and a study or description of the numerous organs of the body I have named, and their functions, *is* the study of Anatomy and Physiology.

Frank.—Why, Father, I thought Anatomy and Physiology was something more than that!

Father.—No, it is just that. But it would be well, perhaps, to express it in a little more exact and scientific manner.

Definitions of terms.

ANATOMY, in its primary meaning, is the art of dissecting, or artificially separating the different parts of an animal body. But the term is more commonly applied to a description of the structure, mechanism and organization of living beings.

PHYSIOLOGY is the science of the functions of all the different parts or organs of animals while in a state of health, or the office which they perform in the economy of the individual.

HYGIENE is the art or science of preserving health, upon which I shall speak at some length in the closing conversations of the series.

All bodies or substances are either organic or inorganic. Organic bodies have organs, upon whose action their growth depends. Animals and plants are included in this division. Inorganic bodies are devoid, both of organs and of life. Earths and minerals are of this class.

All organized bodies possess a power which we call vitality, by which they are developed from within by a process called nutrition, which consists in converting certain substances to their own nature by the action of their organs.

All animal bodies derive their origin from parents, or pre-existing bodies of the same kind, and have a period of growth, maturity and decay, and a limited period of life, which varies with different species, from a day to a century.

Animal or organic life is under the control of law, the violation of which shortens the period of life alloted to man; a fact upon which I shall speak more fully in a future conversation.

The various parts of the human body are arranged into what are called *Fil'-a-ments*, *Fi'-bres*, *Tis'-sues*, *Or'-gans*, *Ap-pa-ra'-tus-es*, and *Sys'-tems;* which terms I will explain, or define.

A FILAMENT is a fine thread of which flesh, nerves, skin, etc., are composed.

A FIBRE is a slender thread, composed of several filaments united, and enclosed in a sheath. It constitutes a part of the form of animals.

A TISSUE is formed by a particular union, or interlacing of fibres.

AN ORGAN is an arrangement of tissues forming an instrument designed for action. The action of an organ is called its *function*.

To illustrate. The lungs are organs of respiration, and breathing is one of their functions. The liver is an organ, and its function is the secretion of bile.

AN APPARATUS is a union of several organs for the production of certain results. The digestive apparatus affords a good example. It consists of the teeth, salivary glands, stomach, pancreas, liver, etc., all uniting to aid in the digestion of food.

"The term, SYSTEM, is applied to an assemblage of organs arranged according to some plan, or method; as the nervous system, the respiratory system."

MEM'-BRANE, or TISSUE, is a thin, white, flexible skin, formed by fibres interwoven like a net-work.

There are several kinds of membrane, as the *Mu'-cous, Se'-rous, Cell'-u-lor, Ad'-i-pose, Mus'-cu-lar*, &c.

THE MUCOUS MEMBRANE lines the mouth, nose, stomach, bowels, lungs; in short, all cavities which communicate with the air.

The function of the Mucous Membrane, or the small glands situated in its substance, is to secrete a viscid, slippery fluid called mucous, the object of which is to lubricate the surface, and thereby prevent chafing and irritation.

THE SEROUS MEMBRANE lines all closed cavities which have no communication with the air, such as the skull, chest, etc. It secretes a watery fluid which lubricates the organs or parts it envelopes. This membrane, when diseased, sometimes secretes too great an amount of water, which constitutes dropsy of the brain, chest, and other organs.

Membranes.

The Cellular Membrane is a net-work of cells, varying greatly in size and shape. It is composed of small fibres, running in every possible direction, and extending throughout the entire body. These cells communicate with each other, and the fluid they contain flows from cell to cell, and in consequence of disease, often becomes too great, producing general dropsy.

The Adipose Membrane, or Tissue, forms distinct bags or cells, which contain a substance called *fat*. There is, in many instances, a great deposit of this substance in persons who are corpulent. This material is deposited chiefly around the heart and kidneys, and under the skin and muscles of the abdomen. Persons may become enormously enlarged by an increase of this tissue without any increase in the size of their muscles. "Such a condition," as an interesting writer justly remarks, "is to be deplored; the body having become merely the store-house or depot of myriads of pots of fat."

"The Muscular Membane is composed of many filaments, that unite to form fibres, each of which is enclosed in a delicate layer of cellular membrane. Bundles of these fibres constitute a muscle."

Muscles are designed for hard labor, as I shall explain at a future time. They have much lifting and pulling to do. They are therefore composed of minute fibres.

Mary.—Why are they made of small fibres?

Father.—That they may have the greatest possible strength in the smallest compass.

Fred.—Have they any greater strength for being composed of small fibres?

Father.—Certainly they have. Ropes, you know, are made of small cords or threads, for the reason, that, in no other way can they be made of the same diameter to possess the same strength. The four cables which support the Suspension Bridge at Niagara Falls, are each composed of 3659 small wires, for the reason that no solid iron of the same diameter ($10\frac{1}{4}$ inches) could sustain such an immense weight. So the muscles, made up of vast numbers of hair-like fibres, possess a degree of strength truly astonishing, and almost incredible, as I shall show when describing them.

The NERVOUS TISSUE consists of two substances. The one, is of a gray color, and of a pulpy character. The other is white, and of a fibrous character. Both these substances are united in every part of the nervous system.

"THE FIBROUS TISSUE consists of longitudinal, parallel fibres, which are closely united. These fibres, in some situations, form a thin, dense, strong membrane, like that which lines the internal surface of the skull, or invests the external surface of the bones. In other instances, they form strong, inelastic bands, called *lig'-a-ments*, which bind one bone to another. This tissue also forms *ten'-dons*, (white cords,) by which the muscles are attached to the bones."

The Cartilaginous Tissue, or Cartilage, is smooth, firm, and highly elastic. The ends of the bones which come together to form joints, are covered with it. Its color is a pearly white.

The Osseous, or Bony Tissue, varies in arrangement of matter and density of structure, in different bones, and at different periods of life. The arrangement of bony matter is such as to give a cylindrical form to some bones, while others are disposed of in plates, or are flat. Some bones are dense and compact, others porous and spongy. A space in the center of long bones is filled with a substance called *mar'-row*, the design or use of which has long been a subject of controversy among physiologists.

But I fear, my children, that you are beginning to feel disappointed, and have already come to the conclusion, that after all that has been said, I have introduced to your attention only a dry subject, one in which you can never feel any interest.

Frank.—I have no such feeling, Father. It is true, I feel desirous to know what there is in the formation of our bodies to prove God's goodness. But I know it is necessary that we should have knowledge of the

ANATOMY AND PHYSIOLOGY

The human skeleton.

Fig. 1.

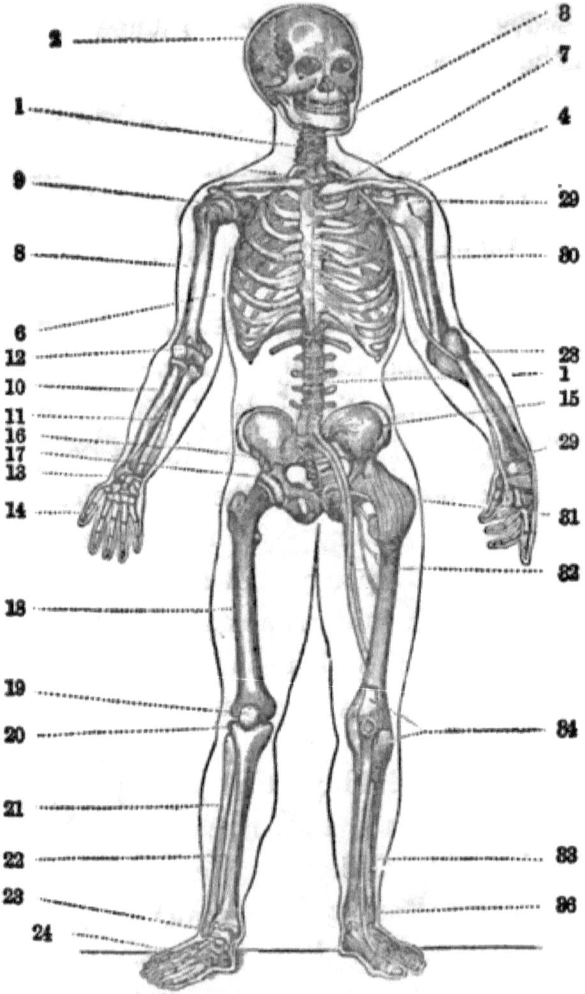

THE HUMAN SKELETON.

Fig. 1. 1, 1, represents the spinal column; 2, the skull; 3, the lower jaw; 4, the breast bone (sternum); 6, the ribs; 7, the collar bone; 8, the bones of the upper arm (humerus); 9, the shoulder joint; 10, the radius; 11, the ulna; 12, the elbow joint; 13, the wrist; 14, the hand; 15, the haunch bone; 16, the sacrum; 17, the hip joint; 18, the thigh bone; 19, the knee pan (patella); 20, the knee joint; 21, the fibula; 22, the tibia; 23, ankle joint; 24, the foot; 27, 28, 29, the ligaments of the shoulder, elbow, and wrist; 30, the large artery of the arm; 31, the ligaments of the hip joint; 32, the large blood vessels of the thigh; 33, the artery of the leg; 34, 35, 36, ligaments of the knee pan, knee, and ankle different parts and structures of which our bodies are composed, else we could not understand your explanations.

Father.—That is true, my son. And I must bespeak your patience a few moments longer while I give a very brief description of the bones of which our bodies are composed, which by Anatomists is called

OSTEOLOGY.

There are, in the human body, two hundred and forty bones, including the teeth, which are usually reckoned at thirty-two.

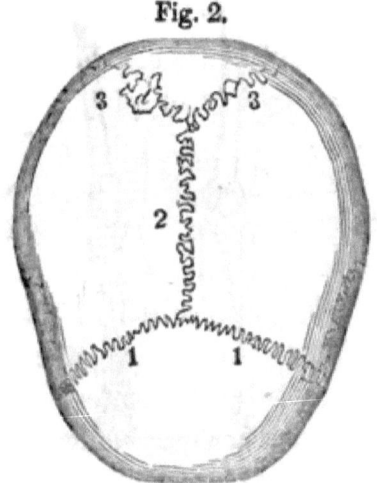

Fig. 2.

THE HEAD.

Fig. 2. 1, 1, the front and upper part of the skull and coronal suture: 2, the sagittal suture on the top of the skull: 3, 3 the lambdoidal sutures running down on each side of the back part of the skull.

Bones of the Head.

Frank.—Why, Father, your first statement surprises me! Can it be possible that there are two hundred and forty bones in our body? I did not suppose there was a tenth part of that number. I should like to know where they all are.

Father.—I will tell you where they are, and will also present to you these fine engravings to illustrate the bones I describe, which I wish you to study carefully. And that you may the better understand, I will divide them into four parts.

1st. The bones of the HEAD, including the teeth, are sixty-two; eight in the *skull*, four in each *ear*, and fourteen in the *face*.

Fred.—I can see but *four* bones in the *skull*.

Father.—The skull bones are double; that is, they are formed of two tablets or plates, one underneath the other, with a porous, spongy layer of bone between. These two plates of bone give very great protection to the brain, the outer one being tough and fibrous, and the other very hard. The bones of the skull are united by what are called *sut'-ures*, which are ragged edges, interlocking each other somewhat in the manner of what by carpenters is styled dovetailing. These bones are quite open in early life, but in old age the sutures close and the bones are firmly united.

The bones of the ears will be described in a future conversation when we come to speak of that organ.

Bones of the Trunk.

The bones of the FACE serve for the attachment of the powerful muscles by which the jaw is moved in masticating food. They also sustain the soft parts of the face in place.

2d. The bones of the TRUNK—fifty-four—twenty-four in the *Spi'-nal Col'-umn*, (back bone); twenty-four *Ribs*, four in the *Pel'-vis;* one at the base of the tongue called the *Os-hy-oid'-es*; and the *Ster'-num*, (breast bone).

These bones are so arranged as to form the two

Fig. 3.

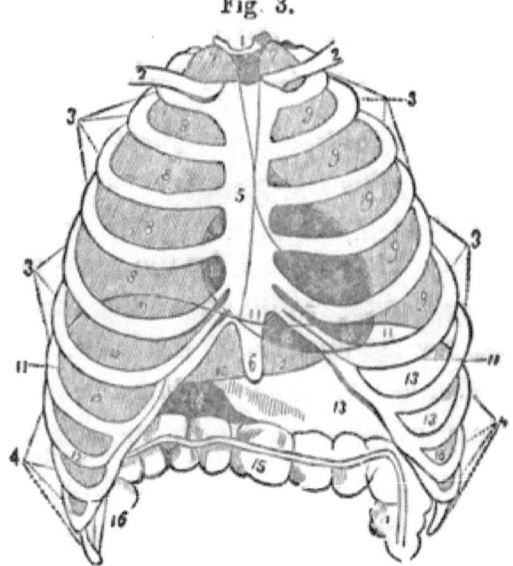

THE THORAX OR CHEST.

Fig. 3. 1, is the spine; 2, 2, the collar bones; 3, 3, the seven upper, or true ribs; 4, 4, the five lower, or false ribs; 5, the breast bone—to which the true ribs are united; 6, the sword-shaped cartilage which constitutes the lower end of the breast bone, called ensiform cartilage; 7, 7, the upper part of the two lungs; 8, 8, the right lung, seen between the ribs; 9, 9, the left lung; 10, 10, the heart; 11, 11, the diaphragm, or midriff; 12, 12, the liver; 13, 13, the stomach; 14, 14, the second stomach, or duodenum; 15, the transverse colon; 16, the upper part of the colon on the right side; 17, upper part of colon on left side.

Bones of the Trunk.

great cavities of the body: the *Tho'-rax* (chest) and the *Ab-do'-men*.

The THORAX is formed by twelve bones of the spinal column posteriorly, or on the back part, the sternum in front, and the ribs at the sides. The sternum in childhood consists of eight pieces of bone, which in adults are so united as to form but three.

The seven upper RIBS, called the *true ribs*, are united in front to the sternum, and in the rear to the spinal column. The next three, called *false ribs*, are united to the spine, and in front are united to each other by cartilage. The other two are connected only with the spine, and are called *floating ribs*.

The SPINAL COLUMN is composed of twenty-four bones, each of which is called a *ver'-te-bra*. Each bone has seven projections called *pro'cess-es*, to four of which, muscles are attached for the purpose of binding the bones together, while the other three serve as attachments for the muscles of the back.

The *vertebra* are so arranged that a tube is formed, in the spinal column, in which the *spinal cord* is placed.

The PELVIS. The four bones which form the pelvis are the two *in-nom-i-na'-ta* (nameless bones), the *sa'-crum*, and the *coc'-cyx*. In childhood, the innominata consists of three bones, which in adults are united and become one. In these bones is a deep cup-like depression, or socket, called the *ac-e-tab'-u-lum*, in which the head of the thigh-bone is placed to form the hip joint.

Fig. 4.

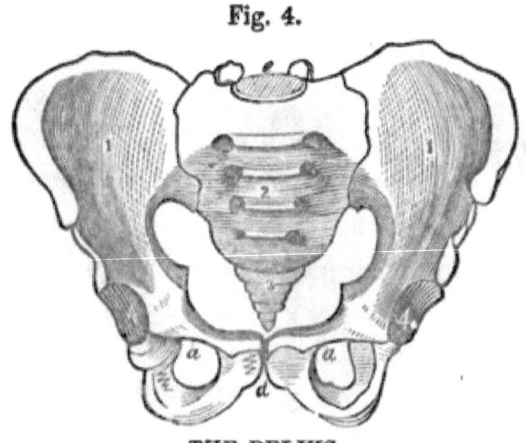

THE PELVIS.

Fig. 4. 1, 1, the innominata: 2, the sacrum: 3, the coccyx: 4, 4, the acetabulum: *a, a*, the pubic portion of the nameless bones: *d*, the arch of the pubis: *e*, the union of the sacrum and the lower end of the spinal column.

The SACRUM. Between the right and left innominata is a wedge-shaped bone called the sacrum, to which the two innominata are firmly bound by ligaments. The sacrum forms the base of the spinal column, to which it is attached by the lower vertebra.

The COCCYX is the lower extremity of the spinal column. In infancy it consists of several small bones, which, in youth unite and become one.

3d. The bones of the *upper extremities*—sixty-four. These are the *Clav'-i-cle*, (collar bone); the *Scap'-u-la*, (shoulder blade); the *Hu'-mer-us*, (first bone of

Bones of the upper extremities.

the arm); the *Rad'-i-us* and *Ul'-na*, (bones of the fore-arm); *Car'-pus*, (wrist); *Met-a-car'-pus*, (the hand); and the *Pha-lan'-ges*, (fingers and thumbs).

The CLAVICLE is attached at one end to the scapula,

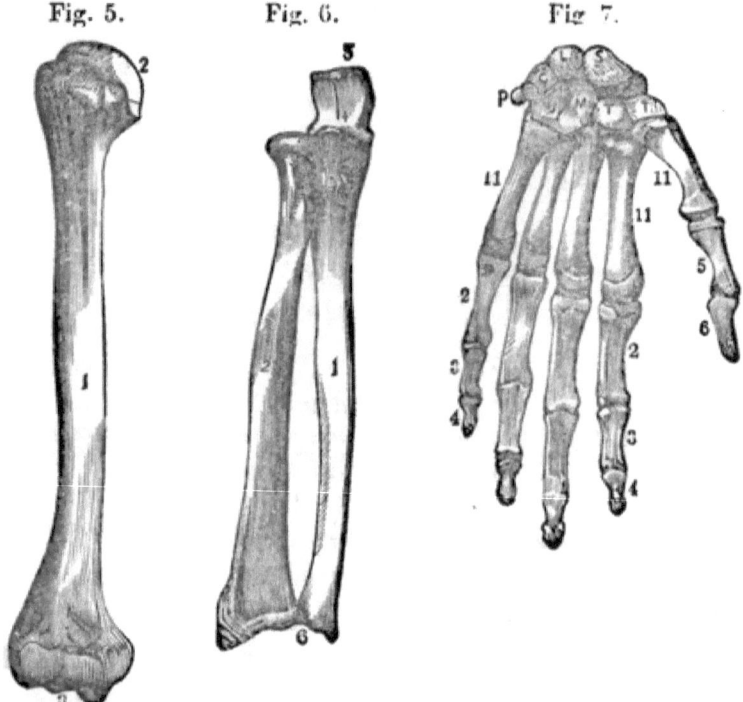

Fig. 5.   Fig. 6.   Fig. 7.

THE HUMERUS.—THE ULNA & RADIUS.

Fig. 5. 1, is the shaft of the bone; 2, the large round head which fits into the glenoid cavity; 3. the surface which unites with the ulna.

Fig. 6. 1, is the body of the ulna; 2, the shaft of the radius; 4, the articulating surface with which the lower end of the humerus unites; 5, the upper extremity of the ulna—which forms the elbow joint; 6, the point where the ulna articulates with the wrist.

Fig. 7. The eight bones of the wrist or carpus, are arranged in two rows, and being bound close together, do not admit of very free motion. S, L, C, P, bones of the first row, T, T, M, U. the second row of the carpal bones, 11, 11, are the metecarpal bones of the hand; 2, 2, the first range of finger bones; 3, 3, the second range of finger bones; 4, 4, the third range of finger bones; 5, 6, the bones of the thumb.

and at the other to the sternum. Its use is to keep the arms in their proper place.

The SCAPULA is a thin, flat bone, of a triangular form. It is located upon the upper and back part of the chest, and is kept in its position by muscles, by the contraction of which, it is moved in various directions.

The HUMERUS is the largest bone of the arm. Its form is cylindrical. The upper extremity is attached to the scapula by the glenoid cavity or socket, forming the shoulder joint. At the other extremity it is connected at the elbow with the ulna of the fore-arm. The fore-arm is the part between the wrist and the elbow.

The ULNA is connected with the humerus and forms the elbow joint. It is situated on the inner side of the arm.

The RADIUS is joined to the bones of the carpus and forms the wrist-joint. It is placed on the outside of the fore-arm.

These two bones of the fore-arm,—the ulna and radius, at their lower extremities, articulate with, or are joined to each other, by which means the hand is allowed to rotate.

The CARPUS is composed of eight bones bound firmly together by strong bands or ligaments.

The METACARPUS consists of five bones. Upon four of them the first bones of the fingers are placed, and the first bone of the thumb upon the other.

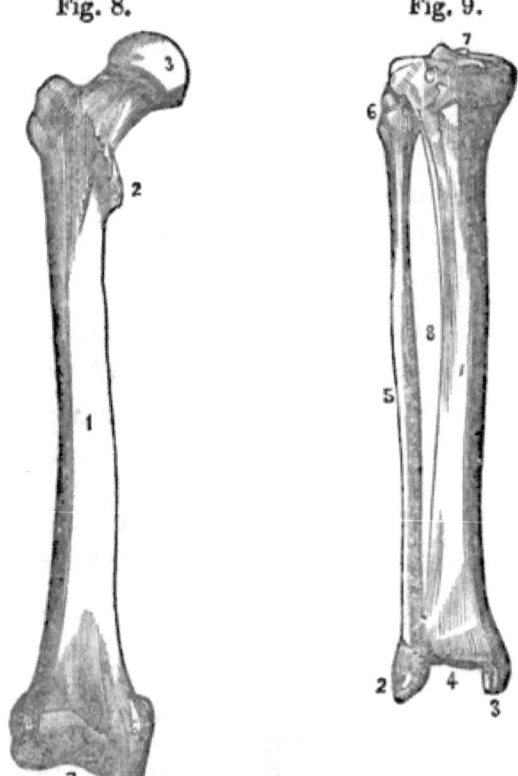

THE FEMUR OR THIGH-BONE.   THE TIBIA AND FIBULA BONES OF THE LEG.

Fig. 8. 1, is the shaft of the thigh-bone (femur); 2, is a projection called the trochanter minor, to which some strong muscles are attached: 3, is the head of the femur which fits into the acetabulum: 5, is the external projection of the femur, called the external condyle: 7, the surface which articulates with the tibia, and on which the patella slides.

Fig. 9. 1, the tibia: 5, the fibula: 8, the space between the two: 6, the junction of the tibia and fibula at the upper extremity: 3, the internal ankle: 4, the lower end of the tibia that unites with one of the tarsal bones to form the ankle joint: 7, the upper end of the tibia which unites with the femur.

The PHALANGES contain twenty-eight bones—three in each finger, and two in each thumb.

4th. The *lower extremities* contain sixty bones; the *Fe'-mur*, (thigh bone); the *Pa-tel'-la*, (knee-pan); the *Tib'-i-a*, (shin bone); the *Fib'-u-la*, (small bone of the leg); the *Tar'-sus*, (instep); the *Met-a-tar'-sus*, (middle of the foot); and the *Pha-lan'-ges*, (toes).

The FEMUR extends from the hip-joint to the knee. It is the largest bone in the body. It has a large,

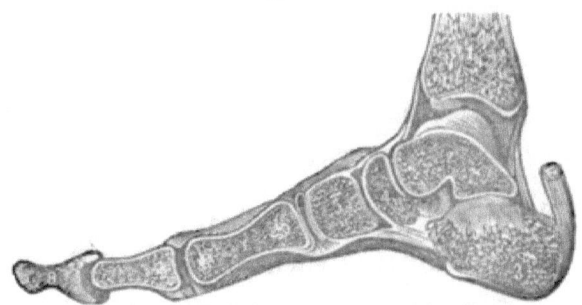

Fig. 10 is a side view of the bones of the foot.

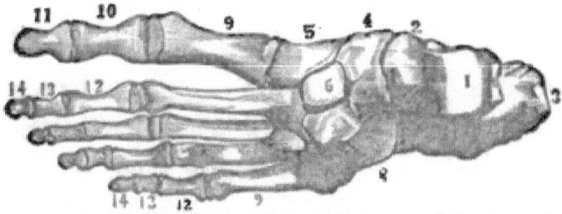

Fig. 11, gives a view of the upper surface of the bones of the foot. 1, is the surface of the as-tra-ga-lus where it unites with the tibia; 2, the body of the astragalus: 3, the heel bone: (os calcis): 4, the sca-phoid bone: 5, 6, 7, the cu-nei-form bones: 8, the cu-poid; 9, 9, 9, the metatarsal bones: 10, the first bone of the great toe: 11, the second bone: 12, 13, 14, three ranges of bones forming the small toes.

round head or ball, fitted into the acetabulum, which as I have said, forms the hip-joint.

The TIBIA and FIBULA extend from the knee to the ankle joint. They are triangular in form, the tibia being the greatest. They are firmly bound together at each end.

The TARSUS consists of seven bones of irregular form, and so firmly bound together as to admit of little or no movement.

The METATARSAL bones are five in number. Their extremities are connected, one with the tarsal bones, and the other with the first bones of the toes. They are so united as to give the foot the form of an arch, concave below and convex above.

The PHALANGES, have fourteen bones; two in each great toe, and three in each small one.

USES OF THE BONES.

The bones, when united, constitute the frame of the system. They are to the body what the timbers are to the house. They retain every part in its proper shape and place, and afford a surface to which the muscles and ligaments are firmly attached, by which the body is bound together, and its motions produced.

The functions of the bones are as various as their form and structure. While some are designed to give protection to the organs they enclose, as the skull, sockets of the eyes, etc., others subserve the purpose of motion, as the upper and lower extremities, while by others, both these objects are accomplished, as the spinal column.

These two hundred and forty bones, like the timbers of a building, must be united or joined together to form the human structure. And such are the varied motions required of the body, that the joining together of the bones must be equally varied. While some require free and unrestrained motion in every direction, others require to be bound fast in their place, or to move within prescribed limits and in particular directions.

A consideration of these facts will lead us to an investigation of the

JOINTS OF THE BODY.

In the first place we observe, that the bones, the connection of which form movable joints, *increase in size* at their extremities or ends, for the purpose, we may suppose, of forming a larger surface for the joint, that the friction may thereby be diminished, and the firmness and stability of the joint increased.

Frank.—I have observed in the bones of animals, that the ends were larger than other parts, but I never before knew why it was so.

Father.—Bone, being composed mostly of phosphate of lime, is of a coarse, rough texture, and consequently, would soon wear out by the friction or rubbing of the joints, had not some effectual means of prevention been devised. And this means is as wonderful as it is useful. It is as follows :—The ends of the bones which come together to form joints, are overlaid with the hard, firm, smooth, and elastic substance called cartilage, (See Fig. 12). The object of this lining is clearly to form a smooth and highly polished surface on which the joint may move without friction or wearing. This lining cartilage is of pearly whiteness, and not more than one-sixteenth of an inch in thickness.

Mary.—Are not other parts of the bones, as well as the ends, covered with cartilage?

Father.—No, my daughter, all other parts are covered with a firm, tough membrane called the *Per-i-os'-te-um*, which adheres very firmly to the bone, and to which the muscles and ligaments are attached, which will be explained hereafter. Nothing can show more clearly the wise design and goodness of the Creator, than this lining cartilage, the object of which —to form a smooth, polished surface for the joint— cannot be mistaken. But *no* surface, not even the *most highly polished steel*, can be made sufficiently smooth to prevent all friction. A constant rubbing for years, will wear away the hardest metals. It is therefore necessary to oil the *most perfect machinery*,

Joints of the body.

to prevent friction; and the same thing is necessary for the joints of our bodies.

Fred.—Do you mean to say that the joints of our bodies need oiling?

Father.—Yes, my son, that is just what I mean.

Fred.—I do not see how that can possibly be done.

Father.—God, in his goodness, has most wonderfully provided for this necessity. On the side of each joint is a gland, which secretes, or separates from the blood, an oily fluid called the *Sy-no'-vi-a*, or joint water, far more slippery and lubricating than the purest oil. A constant supply of this precious mucilage is provided, from the beginning to the close of life, in precisely the quantity required to keep the joints perfectly lubricated or oiled, and so completely is the friction thereby destroyed, that the joints may be in use, without the least wearing, for fifty or a hundred years.

Mary.—O, Father, that is wonderful!

Frank.—I have been told that a steam engine, with its great bars of iron and steel, will wear out in from five to ten years, although it is kept constantly oiled. And will the bones of our bodies wear longer than these large bars of steel?

Father.—Yes, the engine is the work of man, and consequently imperfect. But our bodies are the work of God, and, "as for God, his ways and works are perfect." The joints of our bodies move with such perfect ease and freedom, that we are seldom reminded that we have any joints. Few persons have the slightest idea of the perfect ease with which the joints move. The following facts will illustrate it in a slight degree.

The joint of the knee sustains the weight of nearly the whole body, and in addition to this, a laboring man often carries on his shoulder, a burden of two hundred pounds weight, which, added to that of his body, is not less than three hundred and fifty pounds; and yet the extent of surface of the joint upon which this weight bears, is not, perhaps, more than two square inches. And the movement of the joint is so perfect that not the slightest pain or unpleasant sensation is produced. A man may walk thirty miles a day, which requires not less than thirty thousand steps, and at each step the weight of the body rests upon this small surface of the knee joint, and yet, not the least friction or irritation is produced.

Fred.—How strange, Father, that I never thought of this before. I knew that all kinds of machinery had to be oiled to keep it from wearing, but I never thought that the joints of our bodies needed oiling. How good is our heavenly Father thus to provide for this want.

Father.—This is one of the numerous proofs of the goodness of God, of which we shall discover more and more as we proceed in our investigations.

We will now dismiss the subject and resume it at our next conversation.

QUESTIONS ON CONVERSATION II.

PAGE 24.—What is Anatomy? Physiology? Hygiene? What are all substances? What are organic bodies? What are inorganic bodies? Give an example of each. What do organic bodies possess? From what do animal bodies derive their origin?

PAGE 25. What controls organic life? Into what are the various parts of the human body arranged? What is a filament? a fibre? a tissue? an organ? What is the action of an organ called? Give an illustration. What is an apparatus? Give an example.

PAGE 26.—What is a system? What is a membrane or tissue? Name the different kinds of membrane. What does the mucous membrane line? What is its function? What does the serous membrane line?

PAGE 27.—Describe the cellular membrane. The adipose membrane and its function. Where is the fat chiefly deposited? Describe the muscular membrane. For what are the muscles designed? Why are they made of small fibres?

PAGE 28.—Describe the nervous tissue; the fibrous tissue and its various functions.

PAGE 29.—Describe the cartilaginous tissue, or cartilage. What does it cover? Describe the osseous or bony tissue, and its various arrangements.

PAGE 31.—What is *osteology?* How many bones are there in the human body?

PAGE 32.—How many are there in the head? Describe them.

PAGE 33.—How many in the trunk? Name them. What do these bones form by their arrangement?

PAGE 34.—Of what is the thorax formed? Describe the true ribs; the false ribs; the floating ribs. Of what is the spinal column composed? What are processes, and what is their use? How are the vertebræ arranged? What are the four bones of the pelvis? Describe them and their use.

PAGE 35.—Describe the sacrum and its use. Describe the coccyx. How many bones are there in the upper extremities? Name them.

PAGES 36 AND 37.—Describe the clavicle; the scapula; the humerus; the ulna; the radius. What is the connection between the ulna and radius? How does this affect the hand? Describe the carpus;

ANATOMY AND PHYSIOLOGY.—CONVERSATION II.

PAGE 38.—The metacarpus; the phalanges.

PAGE 39.—How many bones are there in the lower extremities? What are they? Describe the femur;

PAGE 40.—The Tibia and fibula; the tarsus; the metatarsus; the phalanges. What do the bones constitute? What are their uses? What are their functions?

PAGE 41.—In what way are the bones united to form the human structure? What is requisite in regard to their motions? Why are the bones enlarged at the ends?

PAGE 42.—In what way is the wearing of the joints prevented? What is the color and thickness of this cartilage?

PAGE 43.—What additional means is provided to prevent friction? What can you say of the easy movements of the joints? Give examples. Can you discover any proof of goodness of the Creator in these facts?

CONVERSATION III.

JOINTS OF THE BODY CONTINUED.

HINGE JOINT—BALL AND SOCKET JOINT—THE SPINAL COLUMN (BACK BONE)—JOINTS OF THE NECK—PECULIARITIES OF THE KNEE AND ANKLE JOINTS—THE WRIST JOINT—ITS ROTARY MOTIONS—THE GREAT VALUE OF THE HAND.

Father.—I have already intimated, my children, that there are a great variety of joints in the body, some allowing the freest, and others the most limited movements, according to the wants of the different parts of the body.

Fig. 12.

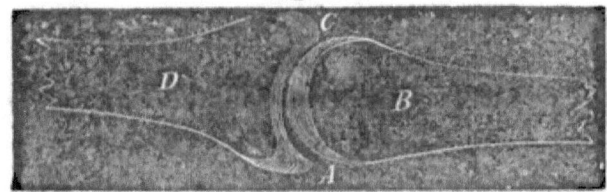

BALL AND SOCKET JOINT.

Fig. 12. D, is the body of a bone, at the end of which is a socket; C, the cartilage lining the socket; B, the body of a bone, at the end of which is a round head; A, the investing cartilage, thin at the sides and thick in the centre.

The most important joints are what Anatomists call the hinge joint, and the ball and socket joint. (See Fig. 12.) The knee and elbow form a perfect example of the hinge joint, and the hip and shoulder of the ball and socket joint.

Frank.—I wish to ask, Father, what keeps the

hinge joints of the knee and elbow together? The hinge of a door, I see, is formed by passing a bolt through the two parts, and by this means, they are kept in their place. Is there any thing like bolts in the hinge joints of our body?

Father.—No, my son, the hinge joints of our bodies are kept in place by a strong, tough membrane, attached firmly to the bones below and above the joint, and enclosing it on every side. This membrane confines and holds the ends of the bones firmly together. The ball and socket joint, also, is kept in place by the membrane I have already described. Yet, as if fears were entertained that this membrane would not prove of sufficient strength to hold the joint together, there is additional security in some important joints, which consists of a strong, firm ligament, one end of which is inserted into the head of the ball, and the other into the bottom of the cavity or cup in which it moves, (See Fig. 13.)

This ligament on the inside, and the membrane on the outside, keep the two parts of the joint in their place with such firmness that nothing but the most unnatural violence can separate them. These joints are perfect examples of mechanical skill, and nothing, perhaps, could present a more striking proof of the wise design and goodness of the Creator.

The important offices and functions of the arms and lower limbs, require that they should move with

Ball and socket joint; its use.

Fig. 13.

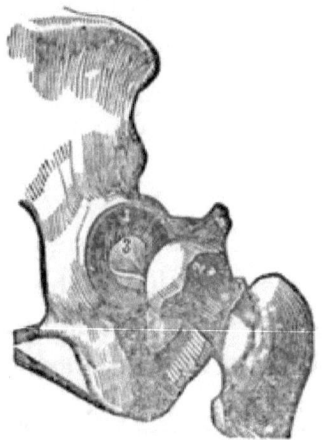

Fig. 13. 1, is the socket of the hip joint; 2, the head of the femur, lodged in the socket; 3, the ligament within the socket.

the utmost freedom in every possible direction. And the ball and socket joint *alone* can allow of, or secure that motion. With this at the shoulder I can move my arm in every direction necessary to perform the varied duties of life ; while the hinge joint at the elbow allows the fore-arm to move in two directions only— backward and forward, and that on a strait line, describing less than one-half a circle.

That we may see more clearly the wisdom and goodness in this arrangement, let us reverse it, and suppose the hinge joint of the elbow had been placed at the shoulder. In that case the arm could only move backward and forward, describing less than half a circle, and consequently would be totally unable to perform the varied motions required.

The spinal column.

Place the ball and socket joint of the shoulder at the elbow, and where would be that steady firmness with which we now move and control the fore-arm and hand? With such an arrangement what worthless appendages to the body would the hands and arms be! Nay more! they would be a great and constant incumbrance, and ever liable to be torn from their sockets by contact with surrounding objects.

Fred.—So ingenious and wonderful is this contrivance I am not surprised that I never thought of this before!

Father.—There is no part of the frame-work of the body, perhaps, that evinces a greater degree of mechanical skill, than the spinal column, (back bone). It is composed, as I have told you, of twenty-four small bones.

Mary.—I would like to ask, Father, why there are so many bones in the spinal column?

Father.—The best answer I can give is, that there should be as many joints as we find really exists. For these twenty-four bones, or vertebras, are joined together by a peculiar and highly elastic substance, the compression or yielding of which at each joint, admits the degree of motion necessary, so that when we bend the body forward or sideways, the elastic substance between each joint is compressed a little on one side, and stretched or lengthened a little on the other, and thus allows the degree of motion necessary

The joints of the neck.

in every direction; the greatest forward, it is true, for in it most is needed. The two first joints of the neck, however, are totally different in their structure and design from any other part of the spinal column. The first, or the one next to the head, is a hinge joint, to allow the backward and forward motions of the head. The other is a mortice and tenon, or ball and socket joint, to allow the *rotary* motion of the head. The hinge joint allows us to look with equal ease, at the starry heavens over our heads, or at the green earth beneath our feet; while the ball and socket joint allows us to turn our head to the right or left; thus enabling us, without a similar motion of the body, to see objects in all directions.

Now let us for a moment, take a view of the results that would have followed, had the joints of the neck been formed like the joints of the spinal column.

In almost every moment of life, during our waking hours, while engaged in the common pursuits of life, we find it necessary to turn the head to the right or left, although we may be unconscious of it. Were it not for the ball and socket joint on which the head rotates, we should be obliged to turn the whole body as often as we now do the head. And what inconvenience and trouble, and even danger, would result from such a necessity. For example: I sit here talking—something calls my attention to the right or left, and instead of turning my head, I must rise from my seat and place my body so as to see it. I am walking in the street, and hear what proves to be a frightened horse attached to a carriage, dashing up behind me. Could I move my head around I should see my danger in time to escape. But as I cannot, I am obliged to stop and turn about, before I can discover my perilous condition, by which time the frantic horse is upon me and crushes me to death.

Now what do you think of this, my dear children?

Fred.—I think when God made us, he saw that our comfort, convenience, and even the safety of our lives depended upon the form of the two joints of the neck, and that he, therefore, in his goodness made them in this way.

Frank.—That is just what I think, Father.

Mary.—And I think so, too.

Father.—A very correct conclusion, my children, I wish now to direct your attention for a moment, to a peculiar and interesting fact in regard to the knee joint. This, as I have before stated, is a perfect hinge joint, but is different in one important respect, from all the others, for this is covered on the front side, by what is called the patella, or knee-pan, which is a curious little bone, the form and use of which are unlike any other. It is circular in form, a little convex or oval on both sides, and is covered on the inside with a smooth cartilage, the same as that with which the joints are furnished. It is kept in place by the

Its form and uses.

powerful tendons which are attached to its upper and lower sides, by which the leg is brought forward in the act of walking. It is not connected with, or attached to any other bone.

Frank.—Of what particular use is this little bone to the knee?

Father.—It serves a number of purposes, the most obvious of which are—

1st. It protects the joint from injury on the side on which it is most exposed.

2d. It serves as a medium of connection between the tendons or ligament by which the leg is moved, one being above, and the other below the knee. But I wish to suggest for your consideration the very significant design of the Creator, in forming this bone and placing it where we find it. Our heavenly Father knew that we should be in duty bound, often to kneel in prayer to him, and it would have been very painful to be, even for a few moments, on the knees, if the knee-joint, like the elbow, had been left unprotected. Let one press the elbow upon a table for a few moments, and see what pain it will cause. But with the knee-joint, covered with the knee-pan, we can remain on our knees for hours with little pain, or inconvenience.

Now I do not say, my children, that this *was* the design of the Creator in forming this bone; I only suggest it for your reflection.

The ankle joint; its form and use.

Frank.—It is certainly worthy of serious reflection.

Father.—Another not less important joint than the knee, is the ankle. This is a hinge joint, but by its peculiar formation, is capable of moving, to a limited extent, like the ball and socket joint; the constantly varying position of the foot requiring such movement. But the most interesting fact in regard to the ankle joint, and that to which I wish to call your attention is, that no joint in the body, perhaps, is so liable to dislocation. In walking and running we are ever liable to place the foot upon an uneven surface, or a rolling pebble, and, with the whole weight of the body upon it, there is great liability of its being turned out, or dislocated. But this is wonderfully guarded against by the peculiar formation of the joint, and what do you suppose it is?

Fred.—I cannot tell; but I have learned to believe, from what you have already told us, that our heavenly Father is both wise and good enough to do any thing that our safety and comfort require. But do explain it; I am becoming so very much interested.

Frank.—I think *I* can tell. I suppose the joint is held more firmly in its place by being surrounded by larger and stouter ligaments than the others.

Father.—That is a very reasonable supposition, my son, but it is not the true cause. What have you to say, Mary? Can you not devise some means for protecting the ankle joint?

Mary.—No, Father, if Frank is not correct, I cannot tell.

Father.—You will recollect that there are two bones between the knee and ankle, called tibia and fibula. On the outside of each, is a projection and prolongation of the outer part, forming what we call the outer and inner ankle bone, the *inner* part of each bone being shorter than the outer. This inner and shorter part of the bones forms the joint, while the outer parts extend down over the joint. Perhaps you do not understand the form of this joint, as it is somewhat difficult to describe. I will therefore use a very simple illustration or explanation. Let a carpenter take a piece of plank, say two inches square, and with an inch chisel cut out the inner or central part of one end, so that each side will be an inch longer than the inside which has been cut out, (See Fig. 9—4). This latter represents the ankle joint, and the two sides which are an inch longer, represent the two bones that project or extend down over the joint on the inner and outer side of the ankle.

Frank.—I think I now understand it. The ankle bone, on each side, extends down lower than the joint, and thereby prevents the joint from slipping out sideways.

Father.—Yes, my son, that is the fact. And why do you suppose the Creator formed the ankle joint so very different from all the others?

Frank.—I can see no cause whatever, but his wonderful goodness.

Father.—There is no other cause. God knew what anguish would be produced by the dislocation of the ankle, and therefore in his goodness formed the joint in such a manner as to render dislocation nearly impossible.

Mary.—I think we ought to love our heavenly Father, when he has manifested so much love and kindness to us.

Frank.—I think so too, and I hope to love him more than I ever did before.

Father.—The more we study the works of God, especially in the formation of our bodies, the greater cause we shall see to love and adore him. But there is one other joint which I must describe, and then I shall proceed to a description of other parts. I refer to the wrist joint, which, perhaps, is the most important and valuable of the body. It is a hinge joint, and in that respect differs but little from others of the same class. It is its *rotary* motion that constitutes its peculiar and great value.

Mary.—What is meant by rotary motion?

Father.—I mean that motion which enables the carpenter to bore with a gimlet, or use a screw-driver; or a lady to use her bodkin. It is, perhaps, not too much to say, that the rotary motion of the wrist is of greater value and importance to man, in the execution of the arts and duties of life, than the motion of all other parts of the body. The hands are the instruments by which the mind works; the mind designs or plans, and the hands execute. But the ability of the hand to execute depends almost entirely upon its rotary motion. From the man who uses the pick-axe and shovel, up through every grade of labor, to the execution of the highest art, the hands alone perform the labor. All other capacities of the body would be comparatively worthless without the hands, and the rotary motion which gives to them their chief value.

Fred.—How amazing are the uses of this rotary motion!

Father.—My son, it is this, more than every other capacity of the body, that distinguishes man from the brute creation. You may think I overrate it, but the truth is, in every thing we do, this motion is necessary, even without our being aware of it. We could not even use either knife and fork, or spoon, without this rotary motion. Indeed, were it not for this, we should be utterly incapable of procuring the necessaries to sustain life. Did you ever think, my children, that to the hand the world is indebted for a practical knowledge of all the arts that adorn life? What would avail the talents of an angel without hands to give practical form to this knowledge.

Frank.—I am surprised people do not speak more frequently than they do of these wonderful facts.

Father.—It is for the reason, I suppose, that like yourself, they "have never thought of them." We must now bring our present lengthy conversation to a close. I hope it has not been altogether uninteresting, or destitute of profit.

Fred.—I must say I have been more interested in this, than any previous conversation.

Father.—Our next subject will be the muscles.

QUESTIONS ON CONVERSATION III.

PAGE 45.—Which are the most important joints? Give examples. How are the hinge joints kept in place?

PAGE 46.—By what means are the ball and socket joints held together? What security additional to the membranes is there in the ball and socket joints? Of what are these joints examples? Of what are they proofs? What do the functions of the limbs require? Describe the difference between a hinge joint and a ball and socket joint.

PAGE 48.—What does the spinal column evince? How are the vertebras joined together? In what way does the elastic substance admit the motions of the spine?

PAGE. 49.—What two joints of the spine are different from the others? Describe them and their important functions. What would have been the results, had the joints of the neck been like the other joints of the spine. Give examples.

PAGE 50.—Describe the patella, or knee-pan, and its various uses.

PAGE 52.—Describe the ankle joint. To what is it peculiarly liable? How is this liability guarded against?

PAGE 53.—What is the peculiar form of the joint, which prevents dislocation? Describe the wrist joint. What gives to it its great value? What is the rotary motion of the wrist? Upon what does the ability of the hand depend?

CONVERSATION IV.

THE MUSCLES.

ALL THE MOVEMENTS OF THE BODY PRODUCED BY THE MUSCLES—THEIR MODE OF ACTION BY CONTRACTION AND RELAXATION—VOLUNTARY AND INVOLUNTARY MUSCLES—THE MUSCLES ENDOWED WITH GREAT POWER.

Father.—In the previous conversations I have spoken of the bones and joints of the human body. But these bones, however wonderful their structure, and perfect the joints by which they are united to form the skeleton, have no power of motion; they cannot move themselves.

Fred.—Do you say, Father, that the bones cannot move themselves? I thought that all the strength of the body was in the bones and joints.

Father.—There is no more capacity or ability in the bones to move themselves, than there is in the timbers of a building; and the joints are designed simply to hold the bones together and allow their movements.

Frank.—If the bones do not move themselves, I should like to know what does move them.

Father.—The instruments which give motion to the animal frame, are the muscles with their tendons.

ANATOMY AND PHYSIOLOGY

Form and use of muscles.

The number of muscles in the human body is estimated at five hundred and twenty-seven. They are made up, as I have said, of small fibres, which, towards the ends, are so modified or changed as to form tendons (cords) and are thereby connected to the surface of the bones, or rather to the firm membrane—the periosteum—which invests, or covers them. The attachment of the muscles to the bone is so firm, that the bone will sooner break than allow the tendon to separate from it. The muscles form an interesting and important part of the system. Also by their large number and size, they constitute the great bulk of the body, and bestow upon it symmetry and beauty. They are situated around the bones of the limbs, so as to invest and defend them, while to some of the joints they form their principal protection. "In the trunk, they are spread out to enclose cavities, and constitute a defensive wall."

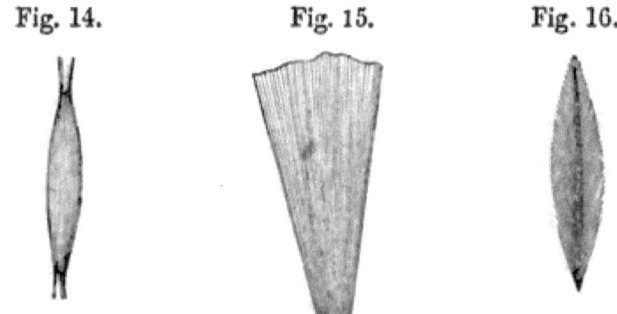

Fig. 14. A spindle shaped muscle.
Fig. 15. A radiate or pen-shaped muscle.
Fig. 16. A penniform muscle, being shaped like the feather end of a pen.

Name and form of muscles.

Muscles vary greatly in form and size. Some are longitudinal and spindle shaped, each extremity terminating in a tendon, while in others the fibres are arranged like the rays of a fan, which converge to a tendinous point and form what is called a *ra'-di-ate* muscle. Again, others are *pen'-ni-form*, which, like the plumes of a pen, or quill, converge to one side of a tendon which runs the whole length of the muscle, or they converge to both sides of the tendon, constituting the *bi-pen'-ni-form*, (See Figs. 14, 15, 16.)

Fig. 17.

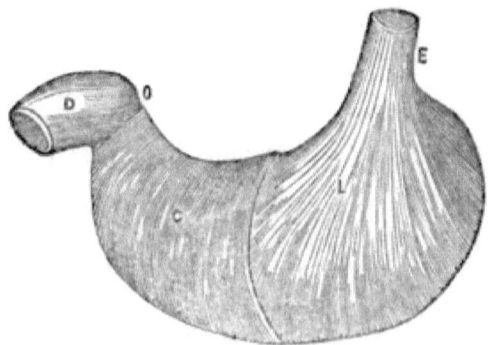

THE STOMACH AND ITS MUSCLES.

Fig. 17. L, represents the fibres running in one direction; C, in another; E, lower end of the gullet; O, pylorous; D, beginning of duodenum, or second stomach.

In describing a muscle we use the terms "*origin*" and "*insertion*." The origin is the part or end firmly attached to a fixed point; the insertion is the part attached to the bone it is designed to move. The middle or fleshy part of a muscle is called the "belly"

How muscles produce motion.

or "swell." Each fibre of a muscle is supplied with both sensitive and motor nerves, also with arteries, veins, and lymphatics.

All the lean or red part of the flesh of man or beast is muscle. You can see the structure of the fibres of which they are composed, by examining a piece of boiled beef. By separating it you will see that it is made up of fine, hair-like threads, each of which is enclosed in a delicate sheath, as I have before stated.

Frank.—If the *bones* cannot move, I should certainly think the muscles cannot, for the bones are a great deal stouter and harder than the muscles, which you say are only the red meat.

Father.—The muscles are endowed by the Creator with the power of contraction and relaxation, which consists in the ability to draw themselves up so as to become shortened, and then to extend themselves to their original length. And all the motions or movements of the body are produced by the alternate contraction and relaxation of the muscles.

Mary.—What do you mean, Father, by alternate contraction and relaxation?

Father.—I mean that the muscle first contracts or draws itself up, and then relaxes, or extends itself. And this alternate action, first contracting and then relaxing, produces each and every movement of which the body, or any part of it, is capable.

Fred.—Do you mean to say Father, that when I

run, the muscles move my legs, or when I raise a weight with my hands, the muscles move both my hands and the weight; that when I talk, the muscles move my tongue and lips, or when I breathe, the muscles move my lungs?

Father.—Yes, Frederick, that is what I mean to say. And it is done, as I have before said, by the contraction and relaxation of the muscles.

Fred.—I do not see how the contraction and relaxation of the muscles can produce these movements. I wish you would explain it, Father.

Father.—I will do so with pleasure. You recollect I told you that towards the ends, the muscles were so arranged as to form tendons or cords, and that these tendons were firmly attached to the bones.

Fred.—Yes, Father, I recollect that, but I cannot see that *that* explains it. I wish that you—

Fig. 18.

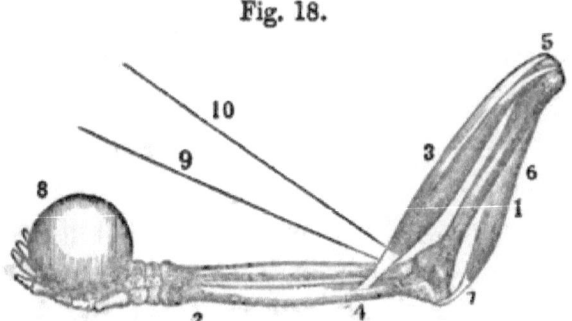

MUSCLES OF THE ARM, AND THEIR MODE OF ACTION.

Fig. 18. 1, is the bone of the arm above the elbow; 2. one of the bones below the elbow; 3, the muscle which bends the elbow; 4, 5, attachments of muscle to bones; 6, the muscle that extends the elbow; 7, attachments to elbow; 8, weight in hand. The muscle, 3, contracts at the central part, or swell, and brings the hand up to 9, 10.

Father.—Be patient, my son, and I will give you a full explanation: I will take the fore-arm and elbow joint, for an illustration of the manner in which the muscles move the limbs and body, (See Fig. 18.)

The tendon of one end of the large muscle on the inside of the arm, is attached to the shoulder bones, where it remains firmly fixed, while the tendon of the other end passes over the elbow joint, and is attached to the bones of the fore-arm an inch or two from the joint.

Fred.—O, Father, I see now how it is. Will you let me explain it?

Father.—Certainly.

Fred.—Well, Father, you say that the tendon of one end of the muscle of the arm is firmly attached to the shoulder bone, so that that cannot move. Now if the muscle contracts and becomes shorter, it must draw up the tendon of the lower end of the muscle which is attached to the bones of the fore-arm, and of course will draw the fore-arm up with it. Is that right?

Father.—That is correct, and is very well explained. The muscle being shortened by contraction draws the bone with it to which it is attached. This is the principle on which all the joints of the body are moved.

Frank.—Does every muscle move a bone when it contracts? You say that there are five hundred and

What produces muscular contraction.

twenty seven muscles, and only two hundred and forty bones for them to move. I do not see what so many muscles have to do.

Father.—Some of the muscles have no connection with the bones. Some organs of the body are composed mostly of muscle, as the heart; and the action produced by the contraction of the muscles of such organs is exerted upon the organ itself. But all the movements of the body, as I have before explained, are produced by the muscles. And as wonderful as it may appear, muscles are provided, and so placed and arranged in different parts of the body, as to produce every conceivable movement of which it is capable.

Mary.—Will you please tell us what causes the muscles to contract, or why a muscle contracts at one time and not at another?

Father.—Muscular contraction is produced by application to the muscle, of its proper stimulus. The muscles that elevate the arm, for example, are called into, or are stimulated to action by the influence of the *will* or mind, and the result is, the arm is raised. Withdraw the stimulus or influence from the muscles, by an effort of the *will*, and they relax and the hand falls.

Mary.—Do the muscles move just because the mind wants them to, and just *where* the mind wishes?

Father.—Yes, my daughter, all the muscles that are under the control of the will.

Voluntary and involuntary muscles.

Mary.—Well, are they not all under the control of the will?

Father.—Why no, my child. Only a part of the muscles are under the control of the will; and they are those that move the arms, fingers, feet, tongue, etc., and are called *voluntary* muscles. Many of the most important muscles of the body act entirely *independent* of the will, under the influence of their appropriate stimulus, as the heart. These are called the *involuntary* muscles. With the *voluntary* muscles we perform all the duties and offices of life; such as walking, talking, writing, the execution of the mechanic arts, indeed all acts of which we are capable. These muscles are incapable of the least action, *only* as they are stimulated or excited by the will.

For example—no one can rise from his seat until he *decides* or *wills* to do so. Nor can he move a foot, or raise a hand, or speak a word, or perform the most trivial act, *only* as the will directs. The mind issues the edict, and the muscles instantly obey the mandate, as if they had heard a sovereign and irresistible call. For example. I wish to raise my hand above my head and instantly my hand goes up. I wish to take something from the floor, and my whole body is instantly bent forward. I wish to look to the right or left, and as quick as thought my head and eyes are turned in that direction.

Fred.—Why Father, how do the muscles know what the mind wishes them to do?

God's goodness manifested.

Father.—The wish or intention of the mind is conveyed to the muscles by, or through the medium of the nerves, which I will explain when we come to converse upon the nervous system. As I have already stated, the will has no control whatever, over the *involuntary* muscles. *They* act and perform their office whether we will or not. Some of these muscles or organs are the heart, the stomach, the bowels, the liver, kidneys, &c. Now mark the goodness of the Creator manifested in this arrangement. The muscles, upon the regular and uniform action of which, life every moment depends, are *placed beyond our control.* Our Maker *knew* that it would not do for us to have the control of them, because we should forget or neglect to keep them in motion or action, and death would be the immediate result. Or, were it possible for us to keep them in motion, *we could attend to nothing else;* nor could we enjoy the repose of sleep for a single moment of our whole life. We should be under the necessity of remaining awake to keep the heart and other organs in motion.

Frank.—How strange, Father, that I never thought of this before! How good our heavenly Father is, to make a part of the muscles and organs of our bodies so that they will act independently.

Mary.—Yes, Father, and how good he was to make the muscles and organs on which *life depends* perform their appropriate duties themselves. The feet, and hands, and head, and tongue and eyes, and many other parts of our bodies, we can move ourselves just when we wish, and it would have been of little consequence if our Creator *had* made these so that they would sometimes go themselves without our care.

Father.—My child, you greatly mistake. Had the muscles which are now under the control of the will been placed *beyond* its control, our life would be in constant and equal danger. We should not be safe for a single moment. We should never know, that we should not the next moment, by the involuntary action of the muscles, be thrown into the fire, or into the water, or under the cars, or into some other place equally fatal to life. Should a man attempt to shave himself, he would be quite as likely to cut his throat as his beard. Indeed, every act of life would be fraught with peril.

Mary.—O, Father, I now see my mistake.

Father.—The muscles are endowed with the power of contracting with great force. The muscles that move the elbow, for example, are capable of an action equal to one thousand pounds weight.

Fred.—That seems to me impossible, Father! I do not see how you can *know* that it is so; certainly you cannot *prove* it, for no one will pretend that he can raise a thousand pounds weight with one hand.

Father.—You should be careful my son, how you make your want of knowledge of a fact, proof that the fact does not exist, or because you see no way to prove it, that therefore it cannot be proved; for the statement I have made is capable of arithmetical demonstration.

Fred.—Well, Father, if you will prove, by the use of figures, that the muscles which move the elbow are capable of acting with the force of a thousand pounds, you will do what appears to me an impossibility, and I will never again doubt your ability to do any thing you may attempt. But I would like to know by what rule in arithmetic you think it can be proved; whether by some of the compound rules, or the rule of three, or single or double position; I think it will be by position, for I recollect, that in solving a problem by that rule, the first thing is to suppose or *guess* at the number, and then proceed upon that supposition, and, if possible, show that you have guessed right. Now is that the way you are going to do, Father?

Father.—No, my son, I shall have no guessing about it. I shall prove it by the rule of three, and as you have made some proficiency in that rule, I hope to make the proof clear.

Fred.—I hope you will, Father, but I do not see how it is possible, for I recollect that in the rule of three, or rule of proportion as it is sometimes called, there are three numbers given to find a fourth, which bears the same proportion to one of the given numbers, that the other two given numbers bear to each other. Am I right, Father?

Power of the muscles demonstrated.

Father.—Yes, you are right in regard to your statements about the rule of three.

Fred.—Well then, I would be glad to know how you will get three numbers to work by, out of one muscle. I am really feeling a great curiosity to know how you will prove the strength of a muscle by the rule of three. It seems to me altogether a new use for that rule.

Father.—I see that you are again making your want of knowledge of a fact, proof that the fact does not exist.

Frank.—It seems to me, Father, just as Frederick has expressed it. I do not see how it is *possible* to prove any thing about the strength of a muscle, by figures. But I know you would not say so if it were not true, nor would you attempt to prove it unless you could do so. Now, Father, do not keep us in the dark about it any longer, but please inform us how you will prove that the muscles of the arm are endowed with such strength.

Fred.—You appear so easy, Father, and unconcerned about your ability to prove what you have stated, that I already begin to fear that I shall be ashamed that I have expressed so much doubt. But do let us know how you will prove it.

Father.—If you will give close attention, my children, I will give you the proof. You recollect I told you that the tendon of the lower end of the muscle of

Power of the muscles demonstrated.

the arm, passes across the elbow joint and is attached to the bones of the fore-arm about one inch from the joint, (See Fig. 18.)

Fred.—Yes, Father, I recollect that, but I do not see how that proves any thing.

Father.—You shall know more about it if you will be attentive. I was going to say that a man's arm from the elbow to the fingers is about eighteen or twenty inches in length. Now if the tendon of the muscle is attached to the bone one inch from the joint, and if from the joint to the fingers is twenty inches, can you tell me how much force must be exerted by the tendon or muscle to raise a pound weight in the hand?

Fred.—Yes, Father, that is easy enough. It would require a force of the muscle equal to twenty pounds.

Father.—Very well, that is correct. Now would it not be an easy thing for a stout laboring man to take hold of the ring of a fifty or sixty pound weight and raise it till the elbow joint is bent at right angle?

Fred.—Why yes, I think I could do that myself.

Father.—Very well. I will now give you a statement of the problem in the rule of three, in which you will find my "three numbers." If it require an action or force of the muscle of the arm equal to twenty pounds weight to raise *one* pound in the hand, how much force will it require to raise *fifty* pounds.

Fred.—Why it will require fifty times as much, or

fifty times twenty. But, Father, I see you mean that *I* shall solve the problem myself, and give the proof which I have been so long waiting for you to give.

Father.—Never mind that my son, if you get the proof. Now will you tell me how many fifty times twenty are?

Fred.—One thousand.

Father.—Well my children, do you now feel satisfied that the strength or power of a muscle can be proved by figures?

Children.—Yes, Father, and in the future we will believe every thing you say, however improbable it may appear.

Father.—I do not intend to say any thing that is not true, but I do not wish you to believe what to you appears impossible or even improbable, *merely* because I assert it. I may not be able in all cases to give as clear proof of the correctness of what I state, as I have in *this;* but I shall always be ready to give you my reasons for believing what I assert.

QUESTIONS ON CONVERSATION IV.

PAGES 57 AND 58.—What gives motion to the body? What is the number of muscles in the human body? Describe a muscle; a tendon, and its connection with the bones. What do the muscles constitute? How are they situated?

PAGE 59.—Describe the forms, and give the names of the muscles. What are the "origin" and "insertion" of a muscle? What is the middle or fleshy part called?

PAGE 60.—With what is each fiber of a muscle supplied? What part of the flesh is muscle? In what way can you see its structure? With what are the muscles endowed? How are all the motions of the body produced?

PAGE 61.—Give an explanation of the manner in which contraction and relaxation of the muscles move the limbs.

PAGE 63.—Of what are some organs composed? How are the muscles placed and arranged? For what purpose? What causes muscular contraction? Give an example.

PAGE 64.—Into what are muscles divided? Describe the voluntary and involuntary muscles, also the cause and manner of their action. Give examples.

PAGE 65.—Through what medium does the mind act upon the muscles? What parts of the body are moved by the voluntary muscles? What by the involuntary? Give examples of the action of each.

PAGE 66.—What would be the result, were the action of each reversed? Give an example. With what force are the muscles endowed that move the elbow joint? How can it be proved that they possess such power?

5

CONVERSATION V.

THE MUSCLES—CONTINUED.

THE FLEXOR AND EXTENSOR MUSCLES—THEIR MANNER OF ACTION IN MOVING THE JOINTS—THE FLEXOR MUSCLES MORE POWERFUL THAN THE EXTENSOR—THE MUSCLES CAPABLE OF ACTING WITH GREAT RAPIDITY—THE PASSIONS AND EMOTIONS EXPRESSED BY THE MUSCLES—SOME MUSCLES PARTLY VOLUNTARY AND PARTLY INVOLUNTARY, OR MIXED.

Father.—In our last conversation I explained the manner in which the muscles move the joints. I think you understand the principle. I wish now to ask a question which either one of you is at liberty to answer. When any joint of the body has been bent by the contraction of the muscle, in what way is it straightened?

Fred.—I cannot tell, unless the muscle, by relaxing, presses it back.

Father.—A muscle has power only in one direction, and that is in drawing itself up by contraction.

Frank.—I suppose that when it relaxes, the limb *falls* back into its place.

Father.—That would depend upon the position of the limb. If the elbow should be bent while the arm was hanging by the side, it would fall back when the muscle relaxed. But how could it straighten itself were the arm raised above the head?

Frank.—I cannot tell.

Father.—Well, Mary, what is your opinion? Can you not contrive some way to straighten the limbs after they have been bent?

Mary.—I do not know that I can tell how it *is* done, but I can say how I think it might be.

Father.—Well, my child, let us hear.

Mary.—I think there might be muscles on the opposite side of the limb, so that when the one which had bent it, relaxes, the one on the other side might contract and draw it back.

Father.—What do you think, Frederick and Frank, of Mary's plan?

Fred.—I think it must be the right one.

Frank.—I think so, too. Is it right?

Father.—Yes, my children, what Mary has suggested, is that which our heavenly Father has adopted. Every joint in our body has two muscles which act in opposition to each other; one relaxing as the other contracts; the one which had bent the joint relaxes, and the one on the opposite side contracting, brings the limb back to its natural position. The muscle which bends the limb is called the *Flexor*, and the one which straightens it, the *Extensor*. These muscles always act in unison with each other; one relaxing while the other contracts. And it is a very interesting fact, that although both are voluntary, yet we cannot, by any effort of the will, cause them to con-

Flexor most powerful.

tract at the same time. Thus kindly has our Creator guarded against the sad confusion that otherwise must have ensued, (See Fig. 18—6, page 61.)

Fred.—Are these two opposite muscles capable of exerting the same degree of force?

Father.—No, there is a vast difference in their power; the flexor which bends the joint, always possess a vast deal more power than the extensor which straightens it. The fingers afford a very striking illustration of this truth, for their principal force or power, being on the inside, is produced by the flexor muscles, while the extensor on the back, with which we straighten them, possess but a slight degree of power.

Frank.—But why, Father, is there so much greater strength on the inside than on the outside of the fingers?

Father.—For the simple reason that the inside is precisely where it is needed. If you attempt to remove a barrel of flour on a wheelbarrow, you will find where the strength of the fingers is required. And as the hand is so important a member of the body, was it not kind in our Creator to place the power of the hand where it was most needed? That we may see more clearly the value of this arrangement and the goodness that dictated it, let us reverse it, and place the power of the inside of the fingers on the outside, and that of the outside on the inside. In that case what a waste would exist on the outside, and deficiency on the inside; and how almost utterly useless the hands would be. By thus examining the subject, every reflecting mind must be forcibly impressed with the wisdom and goodness of the Creator.

Fred.—I think so too, Father; and do you not believe it is very displeasing to God, when he has done so much, and been so careful for our happiness, that we do not see it, and consequently never thank him for it? You know how displeasing it is to us, when we make a great effort to do some kind act to a person, to have them take no notice of it, nor thank us for it.

Father.—Yes, my son, it is so, and I am glad to hear you express yourself in this way. The great Galen, one of the ancient "Fathers of Medicine," was converted from infidelity by the study of Anatomy and Physiology, and wrote an Ode of Praise to his Maker. And I do not see how any one can learn these wonderful things of God's goodness manifested in the formation of our bodies, without being impressed with devout feelings of love and gratitude to Him. One of the direct commands of the Bible is, "Consider the works of God." "Consider,"—study, reflect, meditate upon his works;—as direct a command as, "Thou shalt not kill." And which of all his works, is more proper or profitable to study, than our own wonderful bodily structure. Perhaps in no department of his work, can we see more clearly *what he is, by what he does.*

Frank.—I do not understand, by the explanation you have given, how the muscles can move the *shoulder* joint. Can they, if placed on each side, produce all *its* motions?

Father.—At the shoulder, and all other ball and socket joints, there is a very different arrangement of the muscles and tendons. There are several muscles at the shoulder joint, which are placed in the positions, and draw in the various directions necessary to produce all the motions which the joint will allow.

The eye affords another example of the same wise arrangement. There are six muscles attached to the eye-ball, which are so placed as to pull from different points, and thereby move the eye in every possible direction.

In some parts of the body the muscle is placed at a distance from the part where its action is wanted; the same as a steam engine is sometimes stationed at a distance from the load it is designed to move.

The muscles which move the fingers, are not placed in the palms, or on the back of the hands, but in the arm, extending up as far as the elbow, and are connected with the fingers by long tendons which pass under the ligaments at the wrist and along over the hand to the fingers, and joints of the fingers which they move. The muscles which move the toes are not placed in the feet, but in the calf of the leg.

Fred.—Why are these muscles placed in the arms and legs, and not in the hands and feet?

Capacity of the muscles to be educated.

Father.—I can conceive no other reason for this arrangement than the circumstance that if they had been placed in the hands and feet, they would have rendered them large and clumsy, and destitute of the beautiful proportions which they now present.

The muscles may be trained or educated to act with great precision or exactness. The woodman, for example, will strike with the edge of his axe, any number of times, in the same place. The carpenter and smith will strike with the hammer with equal precision, and the lady will sweep the keys of the piano with great rapidity, touching every key with the utmost exactness.

The muscles are capable also of contracting and relaxing with great rapidity. Fifteen hundred letters may be uttered distinctly in a minute, "the pronunciation of each, requiring both the contraction and relaxation of the same muscles, making three thousand actions in one minute."

In some of the inferior animals and insects, the rapid motion of the muscles far surpasses those of man. Who can conceive of the rapid motion of the wings of the humming bird, which must be several thousand a minute. Yet every movement is produced by the alternate contraction and relaxation of the muscles that move the wings.

Frank.—Is this possible, Father?

Father.—It is not only possible, but it is certain. What else can move the wings, as all the movements of the body are produced by the contraction and relaxation of the muscles.

Frank.—What wonderful perfection this shows in the works of God!

Father.—The *passions* and *emotions* are also expressed by the motion and particular conformation and position of the muscles. Thus one particular position of the muscles of the face will produce a smile, others, scorn, contempt, anger, grief, pleasure, or pain.

Mary.—How wonderful, that the muscles should tell our thoughts and feelings.

Father.—It is, my daughter, and much might be said upon the subject, but as I intend in a future conversation to say more about the face, I will not enlarge here.

I have told you that the muscles are either voluntary, and under the control of the will, or involuntary, and independent of it. There are a few exceptions, the most important of which, perhaps, are the following. Those that move the lungs in the act of respiration or breathing, are neither voluntary or involuntary, absolutely; yet, to a certain extent, *they are both.* The act of breathing goes on every moment of life, without any care or effort of ours, although more than a hundred muscles are employed every time we breathe; yet the *manner* of breathing, as to its being fast or slow, and at regular or irregular intervals, is, within certain limits, *entirely under our control.*

The Creator's goodness evinced

Fred.—I see, Father, that it is so; but I should never think of attaching any great importance to the fact that we are capable of breathing fast or slowly, as we please. Do you think it shows any particular benevolence in the Creator?

Father.—I do. This seems to me one of the clearest proofs of goodness, and, if I may so speak, of thoughtful kindness on his part. He appears to have seen that our convenience and happiness required such an arrangement, and to have stepped aside from his ordinary course, and formed the lungs and their action on a principle differing greatly from most other organs.

Fred.—I cannot see how we can be any happier for being able to breathe fast or slowly, than we should be, if we breathed all the time alike, and at regular intervals. I feel a great deal better when I breathe so, for it is only when I have been running, or taking some other violent exercise, that I breathe so fast.

Father.—Respiration or breathing takes place in an adult, about eighteen times a minute, or once in three and one-third seconds, and in a quiescent and healthy state, at about regular intervals. Now had the movement of the lungs, or manner of breathing, been, like the action of the heart, placed beyond our control, who can estimate the sad results that would have followed.

Frank.—While you and Frederick have been talking about it, I have been trying to think what inconvenience it could cause to be obliged to breathe at regular intervals, and I am as much at a loss as Frederick. I know we can breathe at regular or long or short intervals, as we please, but I never thought there was any particular design in it, or that any sad results would have followed had it been otherwise. You speak as if the effects would have been dreadful; please tell us what they are.

Father.—I will name a few of them. It would have been dangerous to attempt to swallow our food, lest a portion of it would have been drawn into the lungs, producing instant suffocation, if not death.

Frank.—How so?

Father.—You know we cannot swallow and breathe at the same time. And if we had no control over our breathing, we might be obliged to inhale, or take in breath, at the instant of swallowing food, which, of course, would draw it into the lungs. But *now*, having control of the lungs, we can cease breathing while we swallow, and thereby avoid all danger of suffocation. Again, the use of the *voice* in singing, would have been *impossible;* consequently the sweet strains of music from the human voice, never would have charmed our ears, and stirred the deep emotions of the soul; nor should we have been thrilled with the eloquence of the orator, or melted with the pathos of the pulpit. All the tones of the voice would have been measured and monotonous; each sound would have continued less than two seconds, and been succeeded by a pause of equal length to take in breath. For example. With the present endowment or capacity of the lungs, I can repeat the following sentence with one breath :—Honor thy father and thy mother that thy days may be long on the land which the Lord thy God giveth thee. But were I obliged to breathe at equal intervals, I should be necessitated to repeat it in this way :—Honor thy father——and thy mother——that thy days may——be long on the—— land which the Lord——thy God giveth thee.

There could be no such thing as modulation, and the various intonations of the voice, upon which the pleasures of social converse, as well as public discourse, so much depend. In singing, no sound or strain could have continued more than two seconds, and then a pause of equal length would have ensued; whereas now, the strains of the voice may be prolonged almost indefinitely, and the time employed in taking breath is scarcely perceptible. In reading, and in common conversation, instead of giving the sense by proper punctuation and emphasis, all would be broken into short sentences, with a pause between each to take breath, as in the example I have given.

Now, tell me, my dear children; can you see no benevolent design of the Creator, or manifestation of goodness, in the present endowment of the lungs?

Children.—O, Father, you prove every thing you

> Sad results, had the lungs been constructed otherwise.

attempt. We should be stupid indeed, if we did not see the truth, and feel the force of what you say.

Father.—There are other muscles, partly voluntary and partly involuntary, or mixed, as they are called, whose action I shall explain to you at the proper time in future conversations.

QUESTIONS ON CONVERSATION V.

PAGE 71.—How can the muscles straighten the joints, as they have the power to act only by contraction?

PAGE 72.—Describe the flexor and extensor muscles, and their mode of action. What interesting fact is connected with the action of these muscles? Do you discover any proof of kindness in this?

PAGE 73.—Which possess the greater power, the flexor or extensor muscles? Give an example. Why is there greater strength on the inside than the outside of the fingers? Was it not kindness in the Creator to place the power of the hand where it is most needed?

PAGE 74.—What interesting fact is related of Galen? What does the Bible command?

PAGE 75.—Explain how the muscles move the shoulder joint and all other ball and socket joints. What is said of the muscles of the eye? Are muscles ever placed at a distance from the parts they move? Give examples. Why are they so placed?

PAGE 76.—Can the muscles be educated? Give examples. Of what are the muscles capable? Give examples. What is said of some animals and insects?

PAGE 77.—How are the passions and emotions expressed? Give examples. What is said of the muscles that move the lungs?

PAGE 78.—What does this arrangement of the muscles prove? How often does respiration take place?

PAGE 79.—What would be the result, were we obliged to breathe at regular intervals? Give examples. What could there not be? What would have been the effect on singing? on reading and conversation? What would the tones of the voice have been? Give examples. Do the facts given in this conversation prove goodness in the Creator?

CONVERSATION VI.

THE NERVOUS SYSTEM.

The brain—The spinal cord—The cranial nerves—The spinal nerves—Nerves of motion, and nerves of sensation—Their mode of action—Nervous fluid—Interesting facts explained, etc.

Father.—In the previous conversations I have given you a brief description of the bones, and the joints by which they are united to form the human skeleton. I have also described the muscles, both voluntary, involuntary and mixed. You recollect I stated that the bones had no power of motion; that the muscles alone produced all the movements of the body; that they were capable of acting with great power, &c. I am now about to make a statement which you will think a contradiction to what I have before said. It is this—the muscles, in and of themselves, *have no more power of motion than the bones.*

Fred.—I should think your statements contradictory, Father, did I not recollect that you said, that although the muscles are capable of acting with great power, yet they could act only under the influence of their proper stimulus. Now I suppose you mean

Fig. 19.—THE NERVOUS SYSTEM.

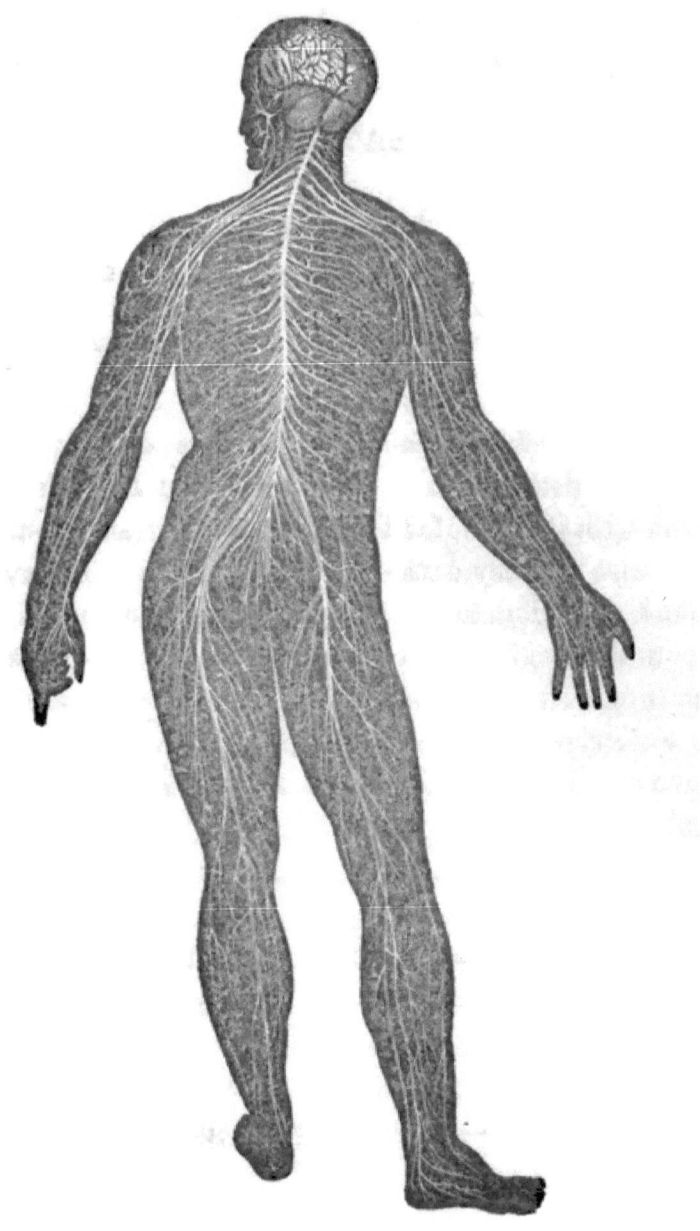

See, hear, feel, taste and smell with the nerves.

that the muscles possess the power, but must be *stimulated* to action in order to act.

Father.—That is correct, and I am pleased that you so readily comprehend my meaning. It is the *nerves* that form the medium of communication between the mind and muscles. Were we destitute of nerves, we should be incapable of motion or sensation, and could neither move, feel, see, hear, taste nor smell.

Fred.—Why, Father, I thought we could see with our eyes, and hear with our ears, and taste with the mouth, and smell with the nose. And do you say it is with the *nerves* we do all this? Why I *know* that I see with my eyes, for I cannot see with them closed, nor hear with my ears stopped. This is quite a new idea! It seems to me that you place a greater value upon the nerves than upon any other, or all other parts of our bodies.

Father.—Do not get excited, my son. All you have said is true, and yet it is true that we see, hear, feel, taste and smell with the nerves. And as to the value I place upon them, you may be assured we can never value them too highly.

Fred.—I have no doubt, Father, that you can explain this, as you have before explained what to me appeared contradictory. But I really feel greatly interested to have you do so.

Father.—It will afford me great pleasure to gratify

your wishes, and at the same time give you instruction in so interesting a branch of knowledge.

I will now proceed to explain, as far as the subject of our conversation requires, the structure and functions of the nervous system.

Fig. 20.

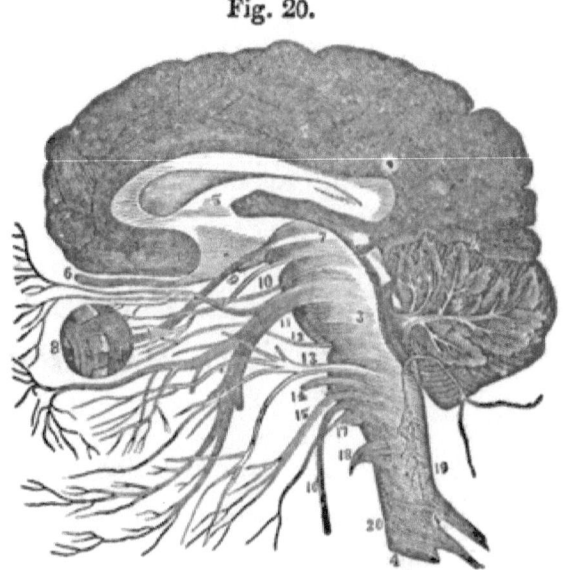

THE CRANIAL NERVES.

Fig. 20. 6, first pair; 7, second pair; 9, third pair; 10, fourth pair; 11, fifth pair; 12, sixth pair; 13. seventh pair; 14, eighth pair; 15, ninth pair; 16, tenth pair; 17, eleventh pair; 18, twelfth pair.

The nervous system consists of the brain, spinal cord or marrow, and forty-three pairs of nerves, twelve of which proceed from the brain, and are called the cranial nerves; and thirty-one from the spinal cord, and are called the spinal nerves.

The BRAIN is a soft, pulpy mass, of a greyish white

Cranial nerves.

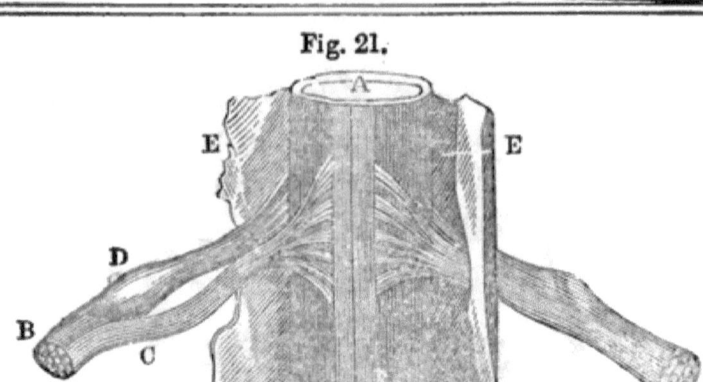

THE SPINAL NERVES.

Fig. 21. Thirty-one pairs connected with the spinal cord. Each pair has two roots, a motor and sensitive; C, motor; arising from the anterior or front columns of the cord; D, the sensitive root springing from the posterior columns. A, is a section of the cord, surrounded by its sheath; B, is the spinal nerve, formed by the union of the motor and sensitive roots. After the union, the nerve, with its motor and sensitive filaments, divides and subdivides as it passes on, and is distributed to the tissues of the several organs.

color, and is contained in the cavity of the cranium or skull.

It is divided into two parts, one called *Cer'-e-brum*, the other *Cer-e-bel'-lum*, or the *great* and *little* brain; the cerebrum being about seven times greater than the cerebellum.

The CEREBRUM occupies all that portion of the skull which lies above the level of the ears, and is the *seat of intellect*.

The CEREBELLUM, or little brain, occupies the lower and back part of the head, and is separated from the cerebrum by a membrane. It is supposed to be the seat of the *animal propensities*.

Cranial nerves.

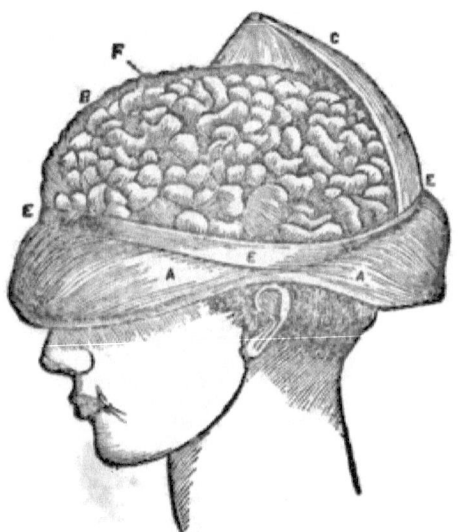

THE BRAIN.

Fig. 22. The skull bones and membranes being removed. The scalp turned down is represented by A, A; E, E, E, show the cut edge of the bones; C, is the dura mata, drawn up with a hook; F, the convolutions of the brain.

There is also the *Me-dul'-la Ob-lon-ga-ta*, which is the top of the spinal cord. It is regarded as a portion of the brain, being enclosed within the skull. It is made up of three pairs of bodies, being so united as to form a bulb.

The brain is invested or covered with a firm membrane called *Du'-ra Ma'-ta*, and also with the *Pi'-a Ma'-ta*, which is a broad net of blood-vessels that extend through the different parts of the brain to supply it with blood. Enveloping the brain, and reflected upon the inner surface of the dura mata, is a serous membrane called the *A-rach'-noid*.

Spinal cord and nerves.

The brain is divided into right and left *hemispheres*, and each hemisphere into three lobes. All the surface of the two hemispheres is divided into *convolutions*, varying in size, and more or less projecting, each of which is supposed to be the seat of some faculty or quality of the mind, (See Fig. 22, page 87).

The convolutions are separated from each other by winding furrows, into which the pia mata descends, while the arachnoid coat, and the dura mata envelope the whole brain.

The *spinal cord* is contained in the cavity of the spinal canal. It consists of both sensitive and motive filaments, distinct in the cord, but united after leaving it, forming *mixed nerves*, which I shall soon explain.

The nerves which proceed from the spinal cord, supply the muscles of at least nine-tenths of the whole body, and are distributed upon the neck, trunk and extremities. They are called the motor and sensitive nerves, or the nerves of motion and sensation.

The cranial nerves, which proceed directly from the brain, are distributed mostly upon the head and upper part of the body; as the optic nerves to the eye, giving vision or sight; the auditory nerves to the ears, giving hearing; the olfactory nerves to the nose, giving the sense of smelling; the gustatory nerves to the mouth, giving the sense of taste.

Fred.—O, Father, I now understand what you mean by seeing, hearing, tasting and smelling with the nerves. You mean that the eyes, ears, etc., are supplied with nerves which give them the power of seeing and hearing.

Father.—That is correct, and I will explain it more fully in future conversations.

The nerves are white cords. They divide and subdivide into branches so minute as to be invisible to the naked eye, and form a network over the whole body, so perfect that the point of the finest needle cannot be introduced under the skin, without touching and wounding one or more. They form a medium of communication between one organ and another, and connect all organs and parts of the body with the brain, which may be regarded as the origin and center of the nervous system.

Frank.—You say, Father, that the spinal nerves are nerves of motion and sensation. Are not the nerves which extend from the brain to the eye, ear, nose and mouth, capable also of feeling or motion?

Father.—No; these nerves can perform no function but the special one for which they were designed; the optic nerves, that of seeing, the auditory nerves hearing, the olfactory nerves smelling, and the gustatory nerves tasting. Should these nerves be touched with the point of a needle, or irritated in any other way, it would cause no pain nor any convulsive motion, as it does in the motive and sensitive nerves. But the effect of the irritation and injury of these nerves, would be to injure or destroy the function of the organ for which they are specially designed—the eyes, ears, etc. Three pairs of the cranial nerves, however, are capable of sensation, and of some degree of motion. The *fifth* pair is capable of very intense pain. It is the sensitive nerve of the face, and is the seat of that painful disease called neuralgia, or *tic douloureux*.

Frank.—I think, Father, we understand your explanation of the action of the nerves which proceed from the brain, and I feel very thankful to you for it; but I do not quite understand about the spinal nerves, which you say are nerves of motion and sensation. I wish you would explain to us how these nerves produce motion and feeling.

Father.—I will do so, my son, with pleasure. You recollect I told you that the nerves were in pairs—that is, a *pair* consists of two distinct nerves, and each nerve of a pair is totally different in its action or function, from the other nerve of the same pair; one being a sensitive nerve, whose function is to produce feeling, the other a motive nerve whose function is to produce motion. The sensitive nerve conveys to the brain and mind, any impression made upon that part of the body upon which the nerve is distributed; and the motive nerve conveys from the brain to the surface or extremities of the body, the intention of the mind; the action of one being from the surface and extremities of the body *inward* to the brain, the

| Action of the nerves. | The nerves a telegraph. |

other *outward* from the brain to the surface and extremities. To illustrate. If I place my hand upon a hot iron, the effect,—the sensation of pain—is conveyed from the hand to the brain and mind, by the sensitive nerve; and the intention of the mind is conveyed from the brain to the hand by the motive nerve. In this case, the sensitive nerve conveys to the brain and mind, the intelligence that something is injuring or destroying the hand. This information is conveyed to the mind, that it may decide what shall be done to save the hand from threatened destruction; and the mind sends back the order to the muscles, *through the motive nerve*, to remove the hand from the iron. The muscles instantly obey the command, and thus the hand is saved.

The *action* of these nerves is with the speed of lightning. If you have any doubts about it, I will tell you how you can remove them, and also prove the truth of what I here state. Just touch a finger to a hot iron, and see how long a time will be required to convey the fact to the brain, and receive back an order to remove it. And yet, all I have here stated of the action of the nerves, must take place, before the finger can be removed from its painful contact with the iron.

This is the first, and most perfect telegraph line ever established; one set of wires (nerves) to convey intelligence from the surface and extremities to the

Action of the nerves.

brain; and another set to carry back the order or intention of the mind, to the surface and extremities. In this way, the condition and wants of every part and organ of the body are made known to the mind. The sensation in the stomach, which we call hunger, which is simply a demand of the system for nutriment, is transmitted to the mind through the sensitive nerves, and the mind sends back to the stomach, through the motive nerves, the order to take food, and obedience to the order at once removes the sensation of hunger. I step upon a pointed nail which pierces the foot, and the fact is instantly transmitted, or telegraphed to the brain, and the brain, as quick as thought, sends back a dispatch, ordering the muscles what to do.

An interesting fact which I must not fail to mention in this connection, is the following:—The sensitive nerve not only conveys to the mind the fact that a part of the system is in danger of being injured or destroyed, but it conveys also the *particular nature* of the injury; or, in other words, the sensation or pain produced by the injury, informs the mind what the injury is. Otherwise, the mind could not act intelligibly in returning an order to the part, what to do to avert the threatened injury.

Fred.—How can that be, Father?

Father.—You know, my son, that different sensations are produced by different causes. If you should place the bare foot upon a hot iron, or live coal, the sensation would be very different from that produced by stepping on ice, or into the snow. In like manner, different diseases, and different locations of disease, produce very different sensations, and thus suggest the proper remedy. The headache is a very different pain from that produced by gout or rheumatism, and that caused by neuralgia is totally unlike that produced by fever. If painful sensations from whatever cause were the same, when the causes producing them are so various, how could the mind be informed of the *true cause?* and if not informed of the cause, how could it direct to the remedy?

Fred.—O, Father! I feel deeply interested in the information you are giving us in regard to the action of the nerves. It is new to me, although I have read something about them. But there is one thing I wish you would explain to us—not that I doubt the truth of it, but I wish to know what proof there is. I refer to the statement you made, that one nerve of each pair is a sensitive nerve, and conveys impressions to the brain; and the other a motive nerve, conveying back the intention of the mind. Now I wish to know how that fact can be proved?

Father.—Your queries are very proper, and I intended to explain the interesting fact, and give the proof, if you had not asked it.

In explaining this, I must first state, that nervous

action is produced by the transmission through the nerve, of what is called the nervous fluid, or nervous force, which some suppose to be identical with, or to strongly resemble the electric fluid. What the exact nature of the nervous fluid is, has not yet been fully decided; but one thing is certain, to prevent or interrupt its circulation, utterly destroys the action of the nerves, both of the sensitive and motive. For example :—If a ligature or thread be tied around the sensitive nerve that is distributed upon the hand, so as to cut off the communication between it and the brain, the sensation of the hand would thereby be entirely destroyed. In that condition, if the hand should be burned off, no sensation or pain would be produced.

If the *motive* nerve should be tied, all power of *motion* would cease. In that case, if the hand should be placed in fire, one would writhe with the pain, but could no more remove the hand than they could remove a mountain. These facts have been proved in numerous instances by experiments on animals.

Mary.—This is wonderful, Father. I am interested and surprised at what you have told us about the nerves.

Father.—My daughter, I have many facts yet to state, in regard to them, not less interesting and impressive than those I have been describing.

I have stated that the spinal nerves were in pairs,

each pair consisting of a sensitive and a motive nerve. These nerves are separate as they leave the spine; one, the motor, proceeding from the anterior or front, and the other, the sensitive, from the posterior or back side of the spinal column, (See Fig. 21, page 86); but they soon unite, and the nervous fibres or filaments of each nerve commingle so as to form, apparently, but one nerve.

Frank.—Well, Father, if they unite and form one nerve, I suppose that nerve can be neither a sensitive nor a motive nerve; and pray, what is it?

Fred.—I think you will find it difficult to answer Frank's question; for I recollect you said that the spinal nerves were the sensitive and motive, and now you say they unite and form one, and that the fibres of each mingle and commingle together. I thought it very fine, when you had two distinct sets of nerves, but now you have mingled them all together, and, it seems to me, have just spoiled the whole, and I do not see what you are going to do with them. I suppose, however, you will find some way to explain, and bring it all right.

Father.—In tracing out and explaining the works of the Creator, we shall never become perplexed, if our explanations and deductions are truthful. In the fact I have stated, and which you seem to think so perplexing, we shall see the wise design and goodness of our heavenly Father. You recollect I told you that the nerves divide and subdivide into branches so minute as to be invisible to the naked eye, and form a network over the whole body.

Fred.—Yes, Father, I recollect that, but I cannot see that that answers Frank's question, or explains the matter in the least. If, after uniting near the spine and forming one nerve, they *do* divide into small branches, I suppose the fibres of each branch are the same as the nerve was before it divided—part sensitive and part motive nerves.

Father.—Yes, Frederick, that is the fact, and in this is to be seen the wise design of the Creator. Each branch, however minute the division, contains fibres of both the sensitive and the motive nerve, and the peculiar, and distinctive quality or character of each nerve *still remains;* that is, the fibres of the sensitive nerve are still sensitive, and those of the motive nerve are still motive, (See Fig. 21, B, and the explanation, page 86). Now with this arrangement, suppose I prick a finger, and touch one of these nerves containing fibres of both the sensitive and motive nerve, the sensitive conveys the fact to the mind; but if there were no motive nerve connected with the sensitive which I had touched, how could the mind convey back its order *to the point from which the sensation or pain proceeded?*

That we may see the more clearly the wisdom and goodness of this arrangement, we will suppose that

Motive and sensitive nerves.

the *sensitive* nerves of one pair is distributed to the *right* hand, and the *motive* nerves of the same pair to the *left* hand. In that case I accidentally place my right hand upon a hot iron. The *sensitive* nerves convey the fact to the mind, and the mind sends back the order to remove the hand; but as the brain or mind is not in communication by the *motive* nerves, with the *right* hand, but is with the left, the order goes off to the *left* hand, and I raise the *left*, but let the right remain on the iron. It is as if one residing in Boston, should receive a telegram from Providence, R. I., and should attempt to return an answer, but by some disarrangement of the wires, the answer goes off to New York.

CONVERSATION VII.

THE NERVOUS SYSTEM—CONTINUED.

SENSATION OF AN ORGAN OR PART, IN PROPORTION TO THE NUMBER OF NERVES DISTRIBUTED UPON IT—THE SENSATION GREATEST WHERE MOST NEEDED—OBJECTIONS ANSWERED—EVERY NERVE A SENTINEL TO GUARD FROM DANGER, ETC.

Father.—At our last conversation I made you acquainted with several interesting facts in regard to the nervous system. There are others not less impressive, which I have reserved for the present interview.

Mary.—You told us so many wonderful things about the nerves, at our last meeting, which I had never known before, that I thought you had said all. But I am pleased to find there are other facts yet to learn; I feel so deeply interested in the subject.

Father.—In those parts of the body where sensation is most needed, the nerves are most abundant; for the sensitiveness of any part or organ is in proportion to the number of nerves distributed upon it.

Frank—What need is there, Father, that one part of the body should be more sensitive than another?

Father.—For the reason, my son, that some parts or organs are used more frequently, and are more im-

Nervous sensation greatest when most needed.

portant in their use than others. For example:—The inside and ends of the fingers are far more sensitive than any other part of the hand, because we need to use the fingers to ascertain the quality or condition of bodies or things, as to their being soft or hard, cold or hot, rough or smooth, etc. And this sensitiveness of the fingers, is precisely where it is needed—where our convenience requires it should be. For who ever thinks of feeling an article with the back of the fingers to ascertain its quality? This sensitiveness is very acute and accurate. By it, the blind may be taught to read, and they may also be taught many of the mechanic arts.

But its great value is seen in the hourly duties of life. The mind reaches out, as it were, through the ends of the fingers and makes itself acquainted with the form and quality of every thing within reach, and appropriates the knowledge thus obtained, to the safety and welfare of the system.

Frank.—Since our last conversation, Father, I have been thinking over what you told us about the sensitive nerves—how they were made to convey painful sensations to the mind, and that all the pain we ever suffer, from whatever cause, is on account of our having sensitive nerves, and that if we had no such nerves we should never have any pain. And you speak of this fact, as proof of the goodness of the Creator. I know that the most you have told us since we commenced conversing upon this subject, has been, to my mind, the clearest proof that God's *goodness* has caused him to make us as we are; but I cannot see how it can be so in this case. If God is so good, I do not see why he could not have made us so that we should never have been liable to suffer pain. O! how delightful that would have been. Then, if I should burn, freeze or bruise my fingers, or flesh, I should have no pain, and if I were *ever so sick* it would cause no suffering. But now, if I burn my fingers, or if I freeze them, I am in such distress for hours that I *don't know what to do*. Or, if I am sick, I have *such* pain in my head, or stomach, or some other part of my body! Or, if I get the smallest mote into my eye, the distress is dreadful.

O, Father, how I wish that I had been made without sensitive nerves. Then, if I should go a little too near a hot stove, it would not burn me, and if I should go into the cold air on the most freezing day, I should feel nothing of it; and no matter how sick I might be, I should have no pain, and even gravel thrown into my eyes, would cause no distress. O, how delightful it would be to live if we could have no suffering!

Sometimes I eat hot bread, or good mince pie, or nice, rich preserves, and it causes me such pain in the stomach, that I am sick for two or three days after. But if I had no sensitive nerves I could eat *just what I like*, and it would never give me pain.

Erroneous views in regard to pain.

Why, it seems strange to me, Father, that you never thought of this! for if you had, I think you would not have said so much about the value of the sensitive nerves, in our last conversation.

Father.—My dear son! you seem not only to have discovered a great improvement upon the works of our heavenly Father, in the formation of our bodies, but you appear very jubilant over your supposed discovery. But do you really think, Frank, that it would have been an improvement in our condition, had we been made so as to be incapable of suffering pain?

Frank.—Why yes, Father, how could it be otherwise? Can any one be as happy while suffering pain, as when free from it? You know you sometimes give what you call opiates to your patients, to relieve their pain. Now, how much better it would have been if they were incapable of feeling pain. I do not know, Father, but you can show that it is a good thing to suffer; but I choose to be free from it, if I can. It is true that you have cleared up some other things that looked dark, and reconciled them with your theory of goodness, but I cannot see how you can do so in this case.

Fred.—I have been listening, Father, to the conversation between you and Frank, and I must say with him, that I cannot see that it proves any goodness in the Creator, to make us capable of so much pain and suffering, when he might have made us incapable of it by just leaving out the sensitive nerves.

Results of the incapacity for nervous sensation.

Mary.—And so it seems to me, Father.

Father.—Well, my dear children! Then you have all given your verdict against me, and not only against me, but against the goodness of our Creator. Now let us examine this subject. You say that in case you had no sensitive nerves, if you should go a little too near a hot stove it would not burn you.

Let us suppose that after being in the cold, you go to a stove to warm yourself; not that you *feel* cold, for you are incapable of feeling, but *supposing* your hands may be cold, you approach the stove—but not knowing whether the heat is moderate or intense—for you cannot feel—you hold out your hands to warm them, and your attention is taken from them by entering into conversation with some one standing by, until you look upon your hands, and to your *utter horror*, you see the *flesh is broiling upon the bones!*

Now, was it a favor to you, in this case, or was it kindness in the Creator, that you were made incapable of feeling either cold or heat?

Again, you say, that if you should go into the cold on a very freezing day, you would feel nothing of it. Let us suppose that on such a day you are skating on the ice, or coasting down the hill on your sled. Being busily engaged in your sport, and not *feeling* the cold, you continue in high glee, until you find that your fingers are stiff—you cannot bend them, and soon the wrist and elbow are stiff; still, you feel no pain. But

Want of nervous sensation and its results.

what is the matter? *Why your hands and arms are frozen stiff*, and when they thaw the flesh will fall off and leave the bones bare.

Now, had you been capable of pain, you would not have remained at your sport in the cold. The painful sensation produced, would have driven you into a warm room, before your limbs had been frozen, and thereby you would have saved them, and perhaps your life.

Again, you say, if I had no sensitive nerves, no matter how sick I might be, I should have no pain. But if sickness caused no distress, we should feel no alarm, and of course should neglect the remedy until the disease became fatal.

You say that the smallest mote in the *eye* causes severe suffering. But did you never think of the fact, that the eye is one of the most important, as well as most delicate organs of the body? that a very slight injury may cause its utter ruin, and that for its safety it was necessary that it should be endued with a high degree of sensitiveness, to enable it to give the alarm of the first and slightest approach of danger? The more precious the treasure, the more carefully should it be guarded. Every pain we feel, from whatever cause, is a friendly warning of danger. It informs us that the organ or part of the body where the pain exists, is invaded by a foe, and is in danger of being destroyed.

Every nerve a sentinel.

You recollect I told you that the nerves form a *perfect network* over the entire surface of the body. Well, every nerve is a sentinel, placed there by God himself, to guard the body from danger and death. Not a particle of the surface of the body as large as the point of a needle remains unguarded. Let a foe approach any part of the body, whether in the form of fire or frost, the sting of a bee, or the cut of a knife, and the sentinels (nerves) instantly convey the fact to the mind; thus, every particle of the body is guarded from the incursions of an unknown foe. In sickness of every form, the pain produced, is a warning that something is destroying the vitality or life of the part where the pain exists, and which must be removed or death will be the sure result. Thus timely warned, the proper medical treatment is applied, and death is averted.

Now tell me, my dear children, whose way do you think best, yours, or the Creator's?

Frank.—O, Father! I *do* wish that I had never said a word about it. It does seem so strange that I could not have looked ahead a little and seen the liabilities you have described. I will never again be guilty of thinking that I am wiser and better than the Creator.

Fred.—I am ashamed, Father, that I gave approval to all that Frank said. I hope I shall some time learn my own ignorance, and be more careful how I expose it.

Nerves most numerous on the surface—the reason why.

Mary.—I did not say so much, as Frank and Frederick, but I see, with shame, that I was just as stupid and wicked as they were.

Father.—We may well be ashamed, my children, when we express, or even indulge a thought, that we can make improvement on the works and ways of God. It is true, that in studying his works, we shall often find what, to us, looks dark and mysterious, and what, perhaps, we shall find difficult to reconcile with our notions of wisdom and goodness. But let it always be a sufficient answer to such thoughts and queries, that we are ignorant, and God is infinite in knowledge, as well as goodness. As we advance in knowledge and goodness ourselves, we shall see more clearly the truth of the inspired declaration—" As for God his ways are perfect."

Fred.—If you please, Father, I would like to ask a question. You say the nerves form a network over the surface of the body. Are they more numerous on the surface than in other parts of the body?

Father.—Yes, my son, very much more abundant on the surface than elsewhere.

Fred.—Why is that?

Father.—Evidently, because the surface is the part of the body principally exposed to injury. No injury from without could be inflicted upon the internal parts, without affecting the external first. It was therefore necessary that the external surface should be the

The sympathetic nerves.

principal seat of sensation, so as to give warning on the first approach of danger, that deep and dangerous wounds, to which all, especially the laboring classes, are always exposed, might be avoided.

Besides the nerves I have described, there is another important system called the *sympathetic* or *ganglionic* system. It consists of a series of *Gan'-gli-a*, or knots, which are small masses of nervous matter, and are situated on each side of the spinal column. They are connected with both the spinal and cranial nerves. The ganglia are nervous centres, from which branches pass off and communicate with other ganglia situated in different parts of the body. The head, neck, and trunk, receive branches from the sympathetic, which control the circulation of the blood, the action of the arteries, veins, capillaries, lymphatics,—the function of absorption and secretion,—indeed, most of the functions of the internal organs, are carried on by these nerves, without the knowledge or control of the mind. By these nerves, all the organs of the body are connected, and such is the power of nervous sympathy that a healthy organ often becomes diseased, in consequence of its connection with a diseased organ by the sympathetic nerves.

Thus, my dear children, we have taken a very brief and rapid view of the nervous system, by which we see that it is the nerves which connect the mind with the external world, and give us all the knowledge we have of the universe around us. Had we no nerves, we could never see the light of the sun, but total darkness would brood upon us forever. Nor should we ever hear the sweet music of the human voice, or the varied harmonies of nature. Nor should we ever have tasted the delicious products of earth, which God has provided in such profusion.

All the knowledge we have of the earth around us, or of the heavens above us, is obtained through the nerves. Were it *possible* for the powers of the human mind to be what they now are, without the nerves, we could know nothing—either of God or his works. Shut out from the universe of God, we could have no conception of anything but our own isolated self, and that would be a consciousness of little worth.

Fred.—We feel very grateful, Father, for the knowledge you have given us about the nervous system, and we have passed you a vote of thanks.

QUESTIONS ON CONVERSATION VII.

PAGE 98.—Where are the nerves most abundant? What is said of the sensitiveness of any part or organ? What need is there that some parts should be more sensitive than others?

PAGE 99.—Give an example. What more is said of the sensitiveness of the fingers? What may the blind be taught by it? In what may its greatest value be seen? What does the mind do?

PAGE 101.—Would it have been an improvement in our condition, had we been made incapable of suffering pain? Give the examples and illustrations on pages 102, 103, 104.

PAGE 105.—What shall we often find in studying the works of God? What should always be a sufficient answer? What shall we see as we advance in knowledge and goodness? Are the nerves more numerous on the surface than elsewhere? Why is that?

PAGE 106.—What other system of nerves is there? Of what does it consist? With what are they connected? What are the ganglia? What receive branches from the sympathetic? What do these nerves control? What is connected by them? What is said of the power of nervous sympathy?

PAGE 107.—What gives us all the knowledge we have of the universe? What if we had no nerves?

CONVERSATION VIII.

DIGESTION AND NUTRITION.

The digestive organs—The process or function of digestion, how performed—The digestive fluids, saliva and gastric juice—Chyme and chyle, how formed—The lacteals, their function—The thoracic duct—Chyle poured into the venous blood—Review of the subject—Mastication or chewing the food, the first act of the digestive process—The act of swallowing, how performed—The glottis and epiglottis.

Father.—In the previous conversations I have given you a brief description of the bones, muscles, and nerves; or, rather, I have given some particulars in regard to them—for it can hardly be called a description, it is so brief. These constitute the principal bulk of what is called the solids of the body. But as yet I have said nothing of the manner in which the body is formed, or what sustains it in existence.

The body is made up of a great number of organs and parts, differing greatly in their functions; yet all acting harmoniously, and each contributing to form the most wonderful and perfect piece of mechanism ever produced.

But how is this machinery formed? Whence come the materials? And what power moulds them into

How the body is formed.

form? Why is one part solid bone, another muscle, another nerve, another membrane? and all in due proportion.

The question to which we are to give particular attention at the present time, is,—What are the materials out of which our bodies are made, and by what process do they become bone, muscle, nerve, and the great variety of parts which compose the system?

If we enter a cotton mill, we look with pleasure and surprise, at the thousands of spindles and looms, and other machinery, moving with great rapidity and accuracy. But a careful examination will show that every part is made, or formed, by mechanical means, and upon well known mechanical principles. But not so with our bodies. The knife, the plane, the hammer, have never been brought into use in forming this wonderful structure.

Fred.—Why no, we know that our bodies are not made with tools, the same as you would construct a house, or a cotton mill. Our bodies grow.

Father.—But *how* do they grow, my son?

Fred.—Why I don't know *how*, but I know they grow, for boys grow up to be men. And you know that I am a great deal taller than I was when Mary was born. I suppose we grow because we eat good food. You know you have often told us that new milk and bread made us grow finely. I don't think I should grow much if I did not eat. You know that

How the body is formed.

John Scott, who was so sick all last summer, and could not eat anything, did not grow at all, and became very thin.

Father.—It is true, my son, that we live by eating food. But how can food, such as bread and milk, rice, and toast, make bone, muscle, etc. Is there any similarity between bread and bone, or milk and muscle?

Fred.—I cannot see that there is, Father. I never thought of that before. I have always been eating and growing, and always thought that somehow good food made me grow, but I do not know how; I wish you would tell us.

Father.—The food we eat, such as bread, meat, fruits, and vegetables, is converted into living human bodies, by the wonderful process called digestion.

Fred.—But, Father, how can food be converted into living human bodies?

Father.—I do not mean to say that *food, itself,* becomes living bodies, but it contains what is called elementary principles, out of which our bodies are formed.

Fred.—Will you explain to us how it is done?

Father.—I will give you, briefly, the *substance* of what is known about it. But this, like all the works of God, has limits beyond which human investigation cannot go. You must always bear in mind, my children, that when we study the works of the Creator, we see only a few of the plainest and most obvious facts, and that beyond all that can be discovered by the most profound scrutiny, lie mysteries as incomprehensible as the Deity himself. And yet, strange to say, these mysteries are so common that we seldom or never think of them; constantly before our eyes, yet we never see them; in every part of our bodies, but we never contemplate them.

But I was about to say, that there is a set of organs, or an apparatus, in our bodies, the office or function of which is, to produce the changes in our food, of which I have spoken, and without which, food could no more nourish us, and cause us to grow, than sawdust or gravel. They are called the *digestive organs*.

Frank.—Will you please tell us what the digestive organs are?

Father.—They are the Mouth, Teeth, *Sal'-i-va-ry* Glands, *Œ-soph'-a-gus*, (Gullet—passage to the stomach); *Stom'-ach, In-tes'-tines*, (Bowels); *Lac'-te-als, Tho-rac'-ic* Duct, Liver, and the *Pan-cre-as*.

The Mouth is the cavity which contains the organs of taste and the instruments of mastication. It is bounded on each side by the cheeks; in front by the lips; above by the *hard palate*, (roof of the mouth); and the teeth of the upper jaw; below, by the teeth of the lower jaw and the tongue; behind, by what is called the *soft palate*, which may be elevated or depressed at pleasure, so as to close, or leave open, the passage to the stomach.

ANATOMY AND PHYSIOLOGY

The stomach, liver, bowels, etc.

Fig. 23.

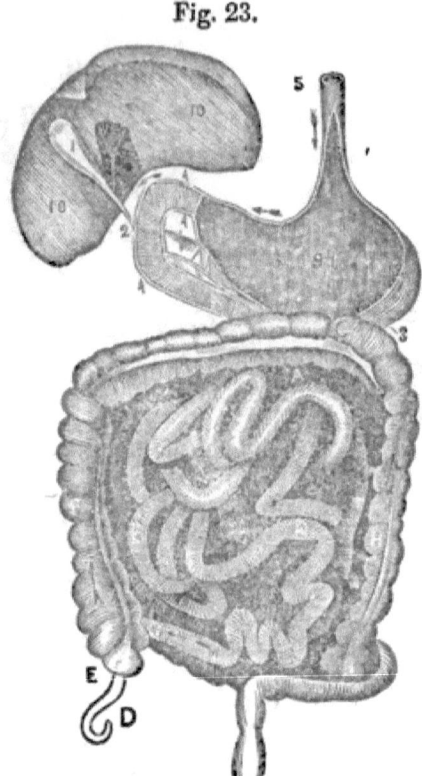

A GENERAL IDEA OF THE STOMACH, LIVER, BOWELS, ETC.

Fig. 23. 9, the stomach; 10, 10, the liver; 1, the gall bladder; 2, the duct which conveys the bile to 4, 4 which is the duodenum; 3. the pancreas; 5, the œsophagus; A, the duodenum; B, the bowels; C, the junction of the small intestines with the colon; D, the appendix vermiformis; E., the cœcum; F, the ascending colon; G, the transverse colon; H, the descending colon; I, the sigmoid flexure; J, the rectum.

As I intend, at some future time, to give you a somewhat lengthy description of the teeth, I shall say nothing here in reference to them.

The salivary glands.

Fig. 24.

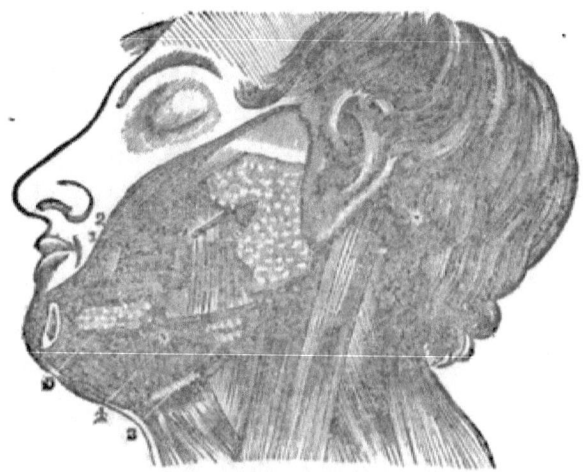

A VIEW OF THE SALIVARY GLANDS IN THEIR PROPER SITUATION.

Fig. 24. 1, the parotid: 2, its duct: 3, the submaxillary gland: 4, its duct. 5, the sublingual gland.

There are six salivary glands—three on each side—called the *pa-rot'-id*, the *sub-max'-il-la-ry*, and the *sub-lin'-gual*. Several small ducts, connected with these glands, open into the mouth, for the passage of the saliva.

The Œsophagus is a large tube extending from the back part of the mouth, or the pharynx, to the stomach. It is situated behind the trachea, heart and lungs. Its function is to convey the food to the stomach.

The Stomach has been compared in form, to a musical instrument called the bag-pipe. It is situated at the upper part of the abdomen, or bowels, a little to the left of the medial line, just below, and in immediate contact with what is called the diaphragm. It has two openings, one for the reception of food, connected with the œsophagus, called the cordiac orifice; the other connects with the duodenum—the first portion of the intestine—and is called the pyloric orifice. These two orifices are at the opposite ends or extremities of the stomach.

The stomach is composed of three coats or membranes. The exterior, is very strong and tough, and by its connection with surrounding parts, keeps the stomach in its proper position. The middle coat is composed of two layers of muscular fibres, one set of which is arranged lengthwise, the other circularly. The inner is called the mucous coat, and is arranged in folds. The stomach is abundantly supplied with small glands which secrete the gastric fluid, which is the great agent, or instrument, in the process of digestion; a description of which, I shall soon give you.

The BOWELS.—Connected with the stomach, is a long tube, called the intestine, or bowels, divided into two parts—the small and large. The small intestine is about twenty-five feet in length, and is divided into three parts, called the *Du-o-de'-num*, the *Je-ju'-num*, and the *Il'-e-um*.

The large intestine is short, compared with the small, being about five feet in length, and like the small, is divided into three portions, namely, the *Cœ'-cum*, the *Co'-lon*, and the *Rec'-tum*.

The small intestine.

Fred.—You surprise me, Father! The intestines *thirty feet long?* What need is there of such length, and how can they be kept in place?

Father.—These are questions of interest, my son, and we will duly consider them, after I have given a description of the different parts.

The DUODENUM is connected with the stomach at the pyloric orifice, as I before stated. It is larger than the other small intestines, and in length, is about the breadth of twelve fingers. It terminates in the jejunum. Ducts from the pancreas and liver open into it about six inches from the stomach.

Frank.—You say the small intestine is divided into three parts. Is there any mark to distinguish the termination of the one, and the commencement of the other?

Father.—There is not. The one is continuous with the other.

The JEJUNUM is continuous with the duodenum. It is somewhat thicker than the other small intestines, and different in color, having a pinkish hue.

The ILEUM is paler than the jejunum—it is also smaller, and thinner in texture. It terminates by a valvular opening into the colon, forming an obtuse angle. By this arrangement, substances are prevented from passing from the colon to the ileum.

How the bowels are kept in place.

The CŒCUM is merely a pouch, a few inches in length, at the commencement of the large intestine.

The COLON is an important part of the large intestine. It is divided into three parts—the ascending, transverse, and descending. It terminates at the rectum, where it makes a curve upon itself, called the *sig'-moid flex'-ure*, which, in the function of the bowels, is of great importance.

The RECTUM is the termination of the large intestine, which, like the smaller, and the stomach, has three coats, or is formed of three membranes.

The mouth, gullet, stomach, and intestine, constitute what is called the *alimentary canal*.

I will now answer your queries in regard to the length of the intestine, and how it is kept in place. I must first state, however, that there is a thick sheet of membrane, called the *mes'-en-ter-y*, formed of several folds of the peritoneum—(lining membrane of the abdomen), spread out from the vertebra, like a fan. The bowels are attached to the edges of this membrane, and held by it in their place. As to the length of the intestine, it does seem a wonderful thing that it should be about six times the length of the body. But a careful study of the subject, will, I think, reveal to us, both the object and necessity of such an arrangement.

Length of the intestine—why so long.

Comparative anatomy reveals the fact, that the length of the intestinal tube, in different species of animals, is governed by the kind or quality of food upon which they subsist; the intestine of the carnivorous, or flesh eating animals, being the shortest, and the herbivorous the longest. In some birds of prey, the intestinal canal is but a little longer than the body of the bird. In the wild-cat it is about three times its length, while the intestine of a sheep is thirty times the length of the body. The longest is in the deer kind, measuring, in some cases, ninety-six feet. But man, being omniverous, living upon a mixed diet composed of both animal and vegetable, holds a medium position, and in him, therefore, we find it of medium length.

Fred.—But why does the kind of food upon which animals subsist, render it necessary that the intestine should vary so greatly in length?

Father.—For the reason that it requires a much greater bulk, and consequently a much longer time to digest, and obtain the nutriment from herbs and grass, than from flesh, and therefore the former require to be retained longer in the alimentary canal, to perfect the digestion, and allow time for the absorption of chyle by the lacteals.

But, when the food is readily digested, and the chyle or nutriment is soon elaborated and made ready for absorption, as animal food is, its long retention is unnecessary, and might be injurious; consequently a long alimentary canal is not needed.

Fred.—I think I understand it now. But I cannot imagine how the small intestine, of such length, can be kept in place.

Father.—It would seem a difficult matter, as it is a soft substance, yielding readily to pressure; some portions of which are laid in folds from side to side, others in oblique and circuitous directions; it must, without some extra precautionary measure for its safety, be constantly liable to displacement by the numerous, violent, and abrupt motions of the body; or be bruised and wounded by every fall; or it would become entangled with itself, or shaken out of its proper position; and the order which is necessary to enable it to carry on its important functions.

Now all this danger, so serious in its nature, and which might be fatal in its results, is provided against in the following manner, which is as admirable as it is beneficial.

Throughout its whole length, the intestine is firmly attached to the mesentery, being "gathered on" to it as a ruffle to a garment; it being not more than one-fourth the length of the intestine. The mesentery not only keeps the intestine in its proper place and position, but sustains the numerous small arteries, veins, lymphatics and lacteals which lead to and from every part of the coats and cavity of the intestinal

Functions of the liver.

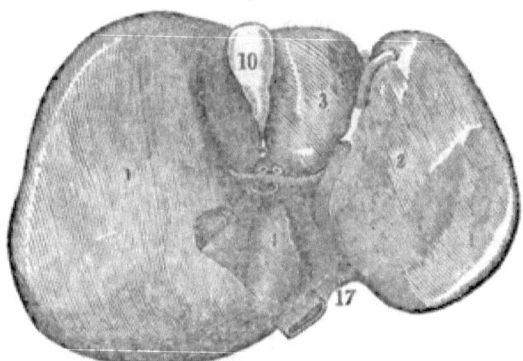

Fig. 25.

THE LIVER.

Fig. 25. 1, the right lobe; 2, left lobe; 3, 4, small lobes; 10, gall bladder; 17, the notch into which the spinal column is fitted.

canal. This membrane, which thus supports the alimentary apparatus, is itself supported by being firmly attached to several of the spinal vertebra.

The LIVER is the largest gland, and, indeed, the largest organ in the system. It weighs about four pounds. It is situated below the diaphragm, in the right side, and has several lobes. It is concave on the under, and convex on the upper surface, and is retained in its place by several ligaments. Its office or function, is two-fold,—the secretion of a fluid called bile—important in the digestive process,—and the separation of impurities from the venous blood. There is a membranous sac, called the *gall cyst*, or bladder, on the under side of the liver, which acts as a reservoir for the bile. A duct, from it, through which the bile is discharged, opens into the duodenum.

Fig. 26.

THE PANCREAS.

"The PANCREAS is a long, flattened gland, analogous to the salivary glands. It is about six inches in length, weighs three or four ounces, and is situated transversely, or across the posterior wall of the abdomen, behind the stomach."

Like the liver, it is connected to the duodenum by a duct, whose mouth opens on its inner surface, through which the pancreatic juice flows.

The OMENTUM, (call), is formed of four layers of serous membrane, connected with the stomach and transverse colon. It contains adipose matter. Its function is to protect the intestines from cold, and facilitate their movements upon each other.

The SPLEEN, (milt), is an oblong, flattened organ, of a dark, bluish color, situated in the left side, under the diaphragm, and in contact with the stomach and pancreas. It was supposed, by the ancients, to be the seat of melancholy. What function it performs is not known, but there are various conjectures, in regard to its office in the animal economy.

Process of digestion described.

The DIAPHRAGM, or MIDRIFF, is a muscular membrane, which separates the thorax from the abdomen. The œsophagus, aorta, and vena cava pass through it. Its functions are various; the most important of which, perhaps, is the aid it affords the lungs in the act of respiration or breathing.

The THORACIC DUCT commences in the abdomen, in front of the lower portion of the spine. It passes through the diaphragm, and rises to the neck. It lies in front of the spine, by the side of the œsophagus and aorta,—a location which protects these important vessels from injury. It is formed by a union of the lacteals, and is about the size of a goose-quill. It terminates at the point where the principal vein of the left arm, called the subclavion, unites with a large vein in the neck, called the jugular. At this point, there is a pair of semilunar valves, the object of which is, to prevent the venous blood from flowing into the thoracic duct;—another proof of the kind care of our Creator.

Having given a brief description of the principal organs concerned in the digestive process, I will now describe the function itself.

Digestion, strictly speaking, is that process by which food is converted into blood. It commences with the mastication and insalivation of the food in the mouth.

Frank.—What do you mean, Father, by mastication and insalivation?

Father.—Mastication is the cutting, or grinding of the food by the teeth. The process is necessary in order to prepare the food for the action of the digestive fluids.

Insalivation is the admixture of the food in the mouth, during mastication, with a fluid called saliva. It is the first of the digestive fluids, and is secreted by the salivary glands, which I have already described.

Mary.—What are glands, Father?

Father.—They are bodies or organs situated in different parts of the body, the function of which is to secrete, or separate from the blood, peculiar fluids, some of which fulfil very important offices in the animal economy. But as I intend to give you a more detailed account or description of these organs, I will here only say, that the salivary glands secrete saliva, or spittle, the object of which is to moisten the food in the mouth.

A fact worthy of note is, that the saliva is not secreted with uniform rapidity at all times. While the tongue and jaws are at rest, and while fasting, the quantity secreted is but small. But the action of the jaws in masticating the food, increases the flow, particularly if the food has a decided taste. The secretion of saliva is great also while masticating *dry* food, as a cracker, and little or none is secreted while eating fluids, as broth or milk.

After mastication and insalivation, the food passes

The lacteals and their function.

into the stomach, where it meets with a powerful and penetrating fluid called the Gastric juice. This fluid mixes with, and disolves the food, forming it into a soft, pulpy mass, called *chyme*. It then passes out of the stomach, through the pyloric orifice, into the duodenum. It there meets and mixes with the bile, or gall, which is a yellow, or deep green, and very bitter fluid, secreted by the liver, and poured into the duodenum through a large tube or bile duct, as I have already explained. While in the duodenum the chyme is also mixed with another fluid called the Pancreatic juice, secreted by the pancreatic glands. By the action of these two fluids, the nutritious portions of the food are separated from the innutritious. The nutritious portion is called *Chyle*, and in color and consistence, resembles thin cream.

The LACTEALS. As the chyle passes along the intestine, it is absorbed, or taken up by vessels called *Lac'-te-als*, whose mouths open on the inner surface, or mucous membrane, of the small intestine.

The lacteals pass between the membrane of the mesentery, and through successive ranges of small bodies called *mesenteric glands*. A large number of the minute lacteals enter the first range of these glands, and a less number of larger ones pass from the first to the second range, and after passing through several successive ranges of these glands, diminishing in

Mesenteric gland and lacteals.

Fig. 27.

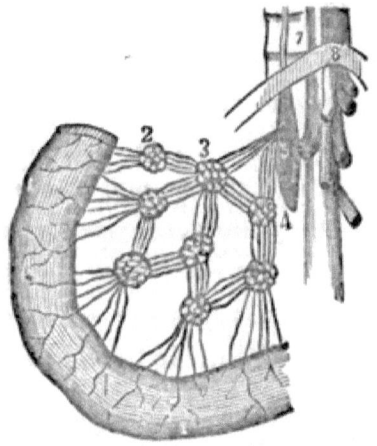

THE MESENTERIC GLANDS, WITH THE LACTEALS PASSING THROUGH THEM.

Fig. 27. 1, is the intestine; 2, 3, 4, the mesenteric glands through which the lacteals pass; 5, the thoracic duct; 7, the spinal column; 8, the diaphragm.

number and increasing in size, they finally enter, and pour their contents into the thoracic duct, (See Fig. 27, which, study carefully).

The lacteals are very numerous, there being millions of them, and are less in size than the smallest hair, being invisible to the naked eye, and can be seen only with the aid of the microscope. They run together, as I have said, and form tubes larger and larger, until they all concentrate in one large tube, the Thoracic Duct, which I have already described. The quantity of chyle taken up by the lacteals in a healthy man, is estimated at from two to three quarts per day.

Frank.—If you please, Father, I wish to ask why

the lacteals are so very small, when the quantity of chyle which passes through them is so great.

Father.—To answer your question, I must first state, that there are millions of blood-vessels in our bodies, so small that three thousand can lie side by side and not extend more than one inch. Now, as the chyle is to be poured into the blood, as I shall hereafter explain, were not the lacteals as minute as these blood-vessels, they might absorb, or take up particles of food so large as to choke up these small blood-vessels, and thus prevent the circulation of the blood, which would cause immediate death. Our Maker knew this, as he made both the lacteals and the blood-vessels, and therefore formed the lacteals so small that they can take up no particles that can obstruct the circulation.

Frank.—Your answer is very interesting, and gives me information of what I knew nothing about before. How clearly it shows the goodness and wisdom of the Creator, and the perfection of his work.

Father.—It does. But I was going to say that the chyle taken up by the lacteals, is poured into the thoracic duct, which conveys it into the veins, where it mingles with the venous blood, and with it is conveyed to the heart, and by the action of the heart and lungs, which I shall hereafter explain, becomes blood, endowed with all the qualities of life.

But this is a very brief description. Let us now review, and enlarge upon what we have been saying upon the subject of digestion, and see if we can discover anything in it that speaks of the goodness and careful kindness of the Creator.

The first act in the digestive process, as we have seen, is the mastication or chewing of the food, so that the digestive fluids can act upon it. This renders it necessary that a suitable apparatus for the purpose, should be provided. And what can be more admirably adapted to the purpose than the teeth, tongue, cheeks, etc. The large, broad teeth, on each side the mouth, do the grinding, while the tongue on the inside, and the cheeks on the outside, serve to keep the food upon, or between the teeth until the grinding is accomplished. At the same time, the movement of the jaw in chewing, stimulates the salivary glands to secrete, and pour into the mouth, a quantity of saliva, for the purpose of moistening the food.

That we may see the value of this apparatus, and the goodness of the Creator in providing it, let us suppose that we are made without teeth, or any thing as a substitute. Without nutriment from food we should soon pine away and die. That we may obtain nutriment from it, it must be digested. That it may be digested, it must be masticated; but how can it be masticated without teeth?

Again, were it possible to eat food without teeth, the great pleasure which people enjoy in eating, would be principally lost; for the rich, delicious taste of food, is obtained mostly by chewing it. But how could we chew without teeth?

Frank.—I think, Father, that I shall value my teeth more than ever, and I trust that I shall be more grateful to my heavenly Father for them.

Father.—I hope you will, and will love him more and more.

Another very interesting fact, is the manner in which food is conveyed from the mouth to the stomach by the act of swallowing. It is as follows:—There are two passages from the mouth to the thorax, or chest,—one for the food, called the œsophegus or

Fig. 28.

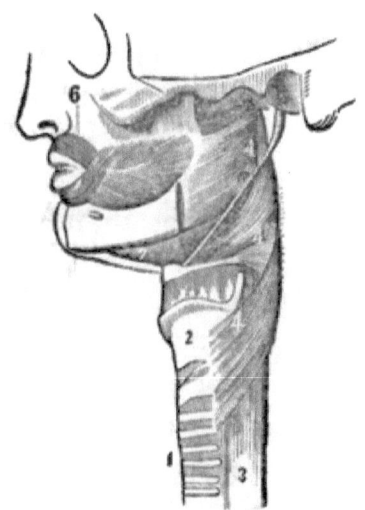

A SIDE VIEW OF THE FACE, ŒSOPHAGUS, AND TRACHEA.

Fig. 28. 1, is the trachea; 2, the larynx; 3, the œsophagus; 4, 4, 4, muscles of the pharynx; 5, the muscles of the cheek; 6, the muscle which surrounds the mouth; 7, the muscle forming the floor of the mouth.

gullet, which I have before described, the other, for the passage of air into the lungs, called the trachea, or windpipe. These passages are so arranged, that the windpipe is placed in front of the gullet,—the passage to the stomach. It is, therefore, necessary, that every particle of food and drop of liquid, that passes from the mouth to the stomach, should, by some means, be conveyed over the upper end, or mouth of the windpipe; and it is also necessary that the windpipe should be kept open to admit the free passage of air into, and from the lungs, in the act of breathing.

Now, how can these two important objects be accomplished, without interfering with each other?

If the windpipe is kept open, how can food and liquids pass over it into the stomach, without going into it? If it is not kept open, how can the air pass freely to and from the lungs?

Fred.—I cannot answer that question, Father, for I cannot think of any way by which it can be done. But I have seen enough, in what you have told us, to assure me, that our Maker can do any thing that is necessary, however impossible it appears to us. Will you please tell us how it is done?

Father.—You are right, my son. God *can* do whatever he pleases, and *will* do whatever is necessary. The plan he has adopted in this case, and the apparatus he has made for its accomplishment, is a wonderful piece of mechanism, which I will attempt to describe.

The epiglottis a safety valve; its mode of action.

The upper end, or mouth of the windpipe, called the *Glot'-tis*, is covered by a cartilage, called the *Ep'-i-glot-tis*, which is a sort of valve, or lid, fitted to the glottis so perfectly, that when closed, it is impossible for a particle of food, or a drop of fluid to pass into the windpipe.

Fred.—It is certainly a good way to have a valve to close over the mouth of the windpipe, but what closes and opens the valve at just the right time?

Father.—This valve is hung, or connected with the parts, by springs made of cartilage, and in its natural condition, is open, and admits the free passage of air to the lungs. These springs are so arranged, that the act of swallowing presses down the valve before the food or fluid gets to it, perfectly closing the passage to the lungs, and leaving the passage to the stomach free and clear, and as soon as the food is passed over the epiglottis, or valve, it opens again of its own accord, and remains open until closed again by the act of swallowing.

This is a *perfectly self-regulating safety valve*, being closed by the same action of the muscles that carry the food into the stomach in the act of swallowing. Now is not this a wonderful arrangement? And just think, this perfectly safe action of the epiglottis, upon which our safety depends, takes place *without the least thought on our part*. Had its action been left to our care, how long do you think we should

Perfect action of the epiglottis.

live? How often we should attempt to swallow, and forget to close the passage to the lungs, and thus produce strangulation, by allowing food to pass into the lungs. Life and death are suspended upon the timely closing and opening of the epiglottis.

Fred.—How impressive this fact! It seems to me, that no one can become acquainted with such proofs of God's kind care for us, without feeling sincere and fervent gratitude.

But, does not the food sometimes get by this valve, into the windpipe?

I have sometimes heard people say that the "food went down the wrong way." And I recollect that once little sister was crying while eating, and some food was drawn down the windpipe with the breath, as mother said; and it caused such coughing and strangling that we feared she would die. Now did not the food go by the valve?

Father.—I think that food never passes into the windpipe in the act of swallowing. But if people talk and laugh, and children cry, with the mouth full of food, particles of it may be drawn in with the breath. But this proves no fault in the structure of the valve, nor does it show any imperfection in its action; it simply shows the imprudence of talking and laughing while eating.

And this leads me to speak of another interesting fact. To draw particles of food, or any other solid

Sensitiveness of the windpipe.

substance into the lungs, endangers life. But our Maker, knowing that we might accidentally, as I have just described, inhale irritating substances, has provided a safe-guard, which is this:—The lining membrane of the windpipe is endowed with the *most acute sensitiveness*, so that the presence of particles of food, or any other substance, causes the most violent coughing and spasm, and thus gives warning of the presence of an enemy that endangers life. And yet this same sensitive membrane is totally insensible to the presence of air, which is constantly passing over it to and from the lungs.

There must certainly be a wise design in this, and it is as good as wise. This sensitive membrane may be regarded as a sentinel, placed there by the Creator, to give instant warning of danger. And how well he performs his duty, every one knows who has had the misfortune to draw a particle of food into the windpipe. But we must now close this conversation which has already been too long, and resume the subject at our next meeting.

QUESTIONS ON CONVERSATION VIII.

Page 108.—What has been described in previous Conversations? What do they constitute? Of what is the body made up?

Page 109.—What is the subject of the present conversation? How do our bodies grow?

Page 110.—How can food make bone and muscle? By what process is food converted into living bodies?

Page 111.—What are the digestive organs? Describe the mouth.

Page 113—The salivary glands; the œsophagus; the stomach. Where is the stomach situated?

Page 114.—What are its two orifices called, and what are their use? Of how many coats is the stomach composed? Describe them. With what is the stomach abundantly supplied? What is connected with the stomach? Into what is the bowels divided? What is the length of the small intestine? Into what is it divided? What is the length of the large intestine? How is it divided?

Page 115.—Describe the duodenum; the jejunum; the ileum.

Page 116.—Describe the cœcum; the colon; the rectum. What constitutes the alimentary canal? Describe the mesentery. How are the bowels kept in their place?

Page 117.—What is revealed by comparative anatomy? What is said of flesh-eating animals? What of herbiverous? What of some birds of prey? What of the wild-cat? What of the sheep? What of the deer? What of man? Why does the food used by different animals require such difference in the length of the intestine?

Page 118.—Name some of the dangers and liabilities of the intestine. How is this provided against, and how is the intestine attached to the mesentery? What else does the mesentery sustain? How is the mesentery itself supported?

Page 119.—Describe the liver. What is its weight? How is its situated? How many lobes has it? What is its form? How is it retained in its place? What is its office? Describe the **gall-cyst** and its use, also its duct.

Page 120.—Describe the pancreas, its length, weight, and situation. To what is it connected, and by what? Describe the omentum and its function. Describe the spleen,— what is said of its function.

ANATOMY AND PHYSIOLOGY.—CONVERSATION VIII.

PAGE 121.—Describe the diaphragm, and its functions. Where does the thoracic duct commence? Through what does it pass? How is it situated? How is it formed? What is its size? Where does it terminate? What is placed at this point? What does this prove? What is digestion? With what does it commence?

PAGE 122.—What is mastication and insalivation? What is saliva? What are glands? What is said of the secretion of saliva? With what does the food mix in the stomach?

PAGE 123.—This fluid does what? What is the food then called? Where does it then pass? Through what? With what does it there meet? Secreted by what? With what other fluid is the chyme mixed in the duodenum? By what is it secreted? What is the effect of these fluids upon the food? What is the nutritious portion called? Describe the lacteals, and their function. Through what do the lacteals pass? What more is said of the lacteals?

PAGE 124.—What is said of the number and size of the lacteals? What quantity of chyle is taken up by them?

PAGE 125.—What fact is stated, and what reason is given why the lacteals are so very minute? Into what do the lacteals pour the chyle? Where is it then conveyed? What does it become?

PAGE 126.—Briefly repeat what has been said upon digestion. What is said of the action of the teeth, tongue, cheeks, jaw, salivary glands, etc. What is said of the value of this apparatus?

PAGE 127.—What other interesting fact is named?

PAGE 128.—What is therefore necessary? How is this accomplished?

PAGE 129.—What is said of the glottis and the epiglottis? How is this valve connected with the parts? What is its natural condition? Describe the action of this valve. What kind of a valve is this? By what is it closed? This action of the epiglottis takes place without what? Had its action been left to our own care, what would have been the result?

PAGE 130.—Does the food ever pass into the windpipe in the act of swallowing? Repeat the substance of what is said on this page.

PAGE 131.—What is said of the lining membrane of the windpipe? As what may this membrane be regarded? Do you discover wisdom and goodness in these facts?

CONVERSATION IX.

DIGESTION AND NUTRITION—CONTINUED.

Quantity of gastric juice secreted—Its power to dissolve food—Peristaltic motion—Office of the bile and pancreatic juice.

Father.—As I have before said, the gastric juice is the great agent or instrument in the process of digestion. This fluid has the power of dissolving food of all kinds with which it comes in contact. It is not found in the stomach at all times, but is secreted at the time of taking food; the food, as it comes into the stomach, acting as a stimulant to the secreting glands. The quantity of gastric juice secreted daily in a man's stomach, is supposed to be not less than fourteen pounds.

Frank.—How can that be? Can the stomach contain fourteen pounds of fluid?

Father.—No, it can hold but two or three pints at a time.

Frank.—What then becomes of fourteen pounds?

Father.—The gastric juice does not accumulate in the stomach in large quantities, but each portion, as soon as it has produced its effect upon the food, is reabsorbed and taken into the circulation, and thus gives room for new secretions of the fluid to continue its action upon the remaining undigested food; and this process goes on until the food is all dissolved, when there will be found but little gastric juice in the stomach, and but little or none will be secreted until food is again taken.

The digestive process is greatly aided by what is called the *per-is-tal'-tic motion* of the stomach. This motion is produced by the alternate contraction and relaxation of muscular fibres running both lengthwise and around the organ. This motion of the stomach brings the food in contact with the mucous membrane, and each portion is mixed with, and is dissolved by the gastric juice, and forms a soft pulpy mass called chyme, as I have before stated. The food is not all dissolved at once, but as fast as it is changed into chyme, it passes out of the stomach, by the pyloric orifice, into the duodenum, and this process continues until the food is all dissolved, after which the stomach rests until another meal is taken.

As I have before stated, the chyme in the duodenum is mixed with the bile from the liver, and the pancreatic juice from the pancreas, and by the action of these fluids, the nutritious portions of the food are separated from the innutritious.

Fred.—How can these fluids produce this effect?

Father.—I cannot tell. It is one of those acts of the Creator which is beyond the comprehension of man. It is known that the gastric juice in the stomach dissolves the food—that the bile and pancreatic juice in the duodenum, separate the nutriment from the gross portions, that the chyle is taken up by the lacteals, and, by a process which I shall hereafter explain, becomes blood. But *how* it is done I cannot tell.

Why these different fluids act with such power upon the food, and with such different results, each producing precisely the effect needed, can be explained only by saying, that the Creator knew these effects must be produced, that our bodies might obtain the nutriment upon which, not only our growth, but our life depends; and, therefore, endowed these organs with the necessary power to secrete the fluids adapted to produce the results I have named. Now all this action of the digestive process which I have been describing, like the action of the epiglottis, and other organs, which I have before explained, is no more under our control, and depends no more upon our will, or any effort we can make, than the rising and setting of the sun. All that we can do, from beginning to end of the process of digestion, is to select, chew and swallow our food. These are voluntary acts of ours. All else is *in*voluntary, and is governed by the laws that control all involuntary action.

And how seldom we think of what is going on within us, and the momentous consequences depending upon the faithful performance of the digestive function.

We take our supper, and retire, and while our bodies are being refreshed with quiet slumber, and our senses are locked in sleep, the digestive organs are busily at work, and before waking in the morning, the nutritious portions of the food taken for supper, are formed into blood, and are coursing through our veins, giving life and vigor to our bodies.

Fred.—I feel very thankful, Father, for the information you have given us upon the subject of digestion; and I think we all have great cause to be thankful to our heavenly Father, that he has so formed the digestive organs, that they do their labor without our care. It seems so strange to me, that I have never seriously thought of these facts before! I wonder if every one is as stupid, in regard to these things, as I have been.

Father.—I do not know, my son, how that is. But from what we see of the community, it would not, perhaps, be harsh judging to say, that most persons are chiefly desirous to get good food—that which is rich and delicious in taste, and after it is swallowed, they think no more of the change that must be produced upon that food by the digestive process, before it can nourish the system,—of the labor that is going

Hunger explained.

on within their own bodies, to prepare that food to nourish and sustain, rather than destroy life, than they think of what is going on in the moon. The most that many seem to think in regard to their food is, whether it has a good flavor, and rich taste; not whether it is such as can be digested, and contains healthy nutriment.

Mary.—I think it is so, Father. But how thankful we should be, that our Maker has so formed the digestive organs, that they perform their duty *whether we think of it or not.*

Father.—There is another fact, in regard to digestion, that I must name. The digestive apparatus not only digests the food, but it gives due notice when the supply is out and more is needed. The sensation that we call hunger, is simply a demand of the system for food. Without it we could never know when our bodies were in real need of nutriment.

How kind it was, then, in our Maker, so to form these organs, that they give timely notice when food is needed. For the appetite would always be a safe guide in regard to both the quantity of food and the time of taking it, had it not been perverted by vicious indulgence.

When food is needed, the stomach sends up a telegram to "head-quarters," giving information that something is needed "down here,"—thus calling attention to the subject.

Facts about digestion.

The mind not only perceives that something is needed, but the telegram (sensation) informs the mind *what* is needed, and food is at once ordered.

Frank.—This is all very interesting and instructive. But I wish to ask if the digestion of food ever causes pain in the stomach?

Father.—No, indeed! Digestion, on the contrary, produces a pleasurable sensation throughout the system. Why did you ask that question, Frank?

Frank.—Because I have heard people complain of pain in the stomach, and I wished to know what caused it.

Father.—When there is pain in the stomach, it is usually caused by indigestion.

Frank.—What causes indigestion, Father. I recollect you told us that the gastric juice had the power to dissolve all kinds of food.

Father.—And so it has. But here is a fact that you should know, and should never forget. The stomach secretes just enough of gastric juice to dissolve the quantity of food the system requires for nourishment, and *no more*. Now, if a man requires but one pound for a meal, and eats two, what will be the result?

Frank.—I suppose one pound will be dissolved, and pass out of the stomach, and the other will remain in the stomach, undigested.

Father.—That is correct. Now, suppose he eats two pounds more at the next meal.

Frank.—He would then have *three* pounds in the stomach.

Father.—When the system required but one, and when there was gastric juice sufficient to dissolve or digest but one. Do you wonder the man had pain in the stomach? And this is the course that thousands pursue, year after year, eating double the quantity of nourishment the body demands, or the digestive fluids can dissolve. Not only is the *quantity* twice what it should be, but the same meal is often composed of a great number of different articles, so compounded as to render them extremely indigestible. But as I intend, at some future time, to speak at length upon the subject of diet and digestion, I will here just add, that the only part God has permitted man to take in the process of digestion,—that of selecting, chewing and swallowing his food, is often so grossly abused, that lingering and painful sickness, and sometimes even death, pays the penalty of the abuse.

Frank.—But what becomes of the food in the stomach that remains undigested?

Father.—It either becomes acid or putrid, and passes off in a crude, acrid state, proving a source of irritation to the whole alimentary canal.

And all this, because our Maker, in great kindness,

Pleasure of eating.

has connected the eating of food, with a great degree of pleasure; and man, in base ingratitude, has well nigh forgotten that food was designed for any other purpose than to please the taste.

But we will leave this subject for the present and resume it at a future conversation.

QUESTIONS ON CONVERSATION IX.

PAGE 132.—What is the gastric juice? What has it the power of doing? When is it secreted? What quantity is secreted daily? What is said of its accumulation and reabsorption?

PAGE 133.—What of this process? By what is digestion aided? How is this motion produced, and what is its effect? What farther is said in regard to the food in the stomach?

PAGE 134.—What is known? What else is said on this page in regard to digestion? What is stated on page 135?

PAGE 136.—What does the digestive apparatus do besides digesting food? What is hunger? What is said of the appetite? What does the stomach do when food is needed?

PAGE 137.—What fact should be known and remembered? What course is pursued? What is said on page 138? What is said of man's ingratitude on page 139?

CONVERSATION X.

THE BLOOD-VESSELS AND LUNGS.

The heart—The arteries—The capillaries—The veins—Changes in the color of the blood—The organs of respiration or breathing, description of, function of—Oxygen taken up by the blood—Carbon thrown off by the lungs—Quantity of air inhaled and oxygen absorbed—Quantity of carbon exhaled—Carbon, how generated—Carbon the food of vegetables, etc.

Father.—At our last meeting I made you acquainted with some facts relative to the very interesting process of digestion, commencing with the mastication of the food, and speaking very briefly, of all the changes that take place, until the chyle is poured into, and mixed with the venous blood.

I wish now to say, that this laborious process of digestion of which we have spoken, has done nothing as yet, towards supplying the system with nutriment. Were it to stop here, we should as surely and as quickly die of starvation, as if we had not tasted a morsel of food. All that has been done by the entire process which I have described, is to form the blood. And the blood now contains all the elements of nutrition which existed in the food previous to digestion.

| Vessels for circulating the blood. |

Now, an entirely new, and totally different process must take place, before the body can receive nourishment, which consists in the conversion of blood into bone, cartilage, muscle, nerve, membrane, etc., and also the repairing the waste of the body, which is going on in every part of the system during every moment of life.

That the blood may nourish the whole system, it must be conveyed to every particle and fibre of the entire body. And this requires ducts or channels, in which the blood can flow; and some effectual means of driving it through them, all of which are provided, and which I will proceed to describe.

The apparatus for circulating the blood, consists of four parts, each differing greatly from the other, viz:

1st. The heart;—a large, hollow, muscular organ, which receives the blood from the veins, at one orifice, and drives it out at another, at the opposite side, into the arteries.

2d. The arteries;—tubes which convey the blood from the heart to all the different organs and parts of the body.

3d. The capillaries;—extremely small vessels which intervene between the minute arteries and veins, and form a connection between the two.

4th. The veins;—a set of vessels which collect the blood from the capillaries, and convey it back to the heart.

Organs of respiration.

In the arteries, capillaries, and veins, the blood varies greatly in quality and color. In the arteries it has a bright scarlet color, but in passing through the capillaries, it loses its bright color, and becomes darker, and in the veins it is a deep purple, and in some parts of the body, nearly black.

Fred.—What causes these changes in the color of the blood, Father? I thought that blood was always red.

Father.—That you may understand the cause of the changes which take place in the quality and color of the blood, as it circulates through the system, you must know the structure and function of the respiratory organs; and this I shall make the subject of the present conversation.

The organs of respiration, or breathing, are the *Lungs*, (lights), the *Tra'-che-a*, (windpipe), the *Bron'-chi-a*, (divisions and sub-divisions of the windpipe), and the *Air-cells* at the extremities of the bronchia.

The LUNGS are situated in the chest or thorax. They are conical in form, and are divided into right and left, the heart being situated between them. The right lung is divided into three lobes, and the left into two. The right is larger than the left.

The lungs are separated from each other by a membrane, or partition, called the *Me-di-as-ti'-num*. They are of a pinkish grey color, mottled.

Fig. 29.

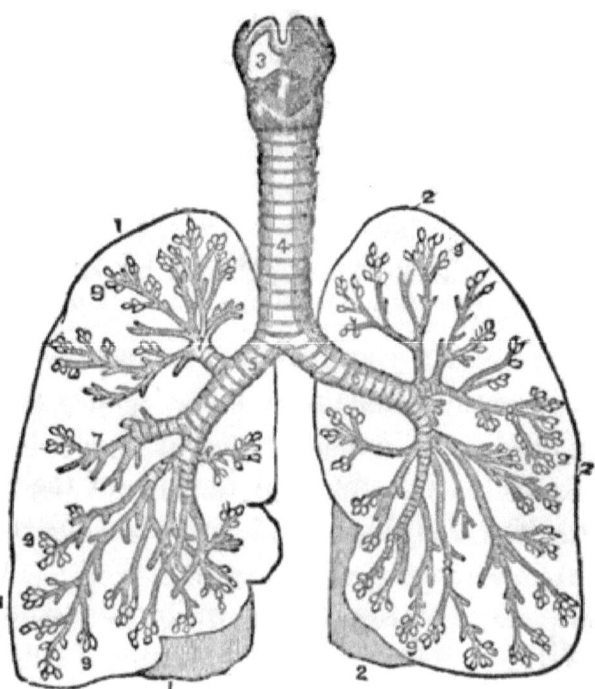

THE LUNGS.

Fig. 29. 1, 1, 1, the right lung; 2, 2, 2, the left lung; 4, the windpipe; 5, the right bronchial tube; 6, the left bronchial tube; 7, small bronchial tubes; 9, 9, 9, 9, 9, 9, air-cells.

Each lung is enclosed with a serous membrane, called the *Pleu'-ra*, which is reflected upon, and lines the chest. When this membrane is inflamed, the disease is called "pleurisy."

The lungs are formed of minute divisions of the bronchial tubes, the air-cells, lymphatics, arteries and veins, connected by cellular tissue, which constitute the *Pa-ren'-chy-ma*.

Air-cells, very numerous.

The lungs are retained in place by the pulmonary arteries and veins, bronchial tubes and vessels, and pulmonary nerves.

The TRACHEA extends from the larynx, (which is, indeed, the upper part of the trachea)—to the third dorsal vertebra. At this point, it divides into two parts, called bronchia. I will describe the structure of the trachea and bronchia, when I explain their functions.

The number of air-cells in the lungs is almost inconceivable. They have been estimated as high as one thousand seven hundred millions. Were the inner surface of these cells spread upon a plain, or level surface, the aggregate, or whole extent, would be more than thirty times greater than the surface of the whole body.

Fred.—Why, Father, you surprise me by such a statement, as you have often done before! But what benefit is such a great number of air-cells, and such an extent of surface?

Father.—This structure of the lungs is as beneficial as it is wonderful, and I will explain it at the right time. But that you may the better understand it, I must first acquaint you with other facts.

The TRACHEA, or WINDPIPE, is made up of rings formed of cartilage, and placed near each other, forming a continuous tube which is kept open by the hard and firm nature of the cartilage. You can feel the hard, bony structure of the windpipe by placing the fingers on the throat.

Now tell me, my children; can you assign any good reason why the windpipe is made of hard, bony rings, and the gullet, which conveys the food to the stomach, is soft and yielding, and is always closed, or collapsed, only when it is opened by the passage of food?

Fred.—I suppose it was made in this fashion that we may breathe and talk easily.

Frank.—I cannot think why it was made so, unless it was that we might laugh heartily.

Mary.—Well, Father, I think it was so that we might sing, clearly.

Father.—Do you not see, if the windpipe was soft and yielding, like the gullet, that when we draw in breath, the sides of the pipe would be drawn in, or would fall together, so as to prevent the air from passing into the lungs?

The windpipe was made of the firm, unyielding cartilage, for the very purpose of keeping the passage for air to the lungs constantly open. Had it been made otherwise, *life would be in danger every moment.* How clearly we can trace in this, the *thoughts of the Creator* when he designed and made the lungs.

Fred.—I never before thought why the windpipe was made so hard. How kind it was of God to make it so that it could not close up when we drew in breath.

Frank.—I never thought any thing about it.

Mary.—Nor I, neither, but I shall, in the future, think *much* about it.

Father.—BRONCHIAL TUBES. The windpipe divides into two branches, called bronchial tubes, one of which passes into the right, and one into the left lung; and these branches continue to divide, and subdivide into tubes, smaller and smaller, until they become invisible to the naked eye, each ending, as before stated, in a minute air-cell.

Fig. 30.

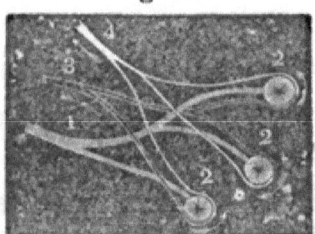

BRONCHIAL TUBES AND AIR-CELLS.

Fig. 30. 1, a bronchial tube, divided into three branches; 2, 2, 2, air-cells; 3, branches of the pulmonary artery winding around the air-cells with the dark blood to be reddened

In the act of breathing, every time we inhale the air, or take in breath, these air-cells are filled with air, which is immediately exhaled, or thrown out of the lungs, and a fresh supply is again inhaled, and then exhaled, and thus the process of breathing goes on from the first to the last moment of life.

Now can you tell me what benefit there is in con-

Oxygen and nitrogen.

stantly drawing air into the lungs, and throwing it out again, or, in other words, in constantly breathing?

Frank.—We *could not live* without breathing!

Father.—But *why* could we not live without breathing?

Frank.—I do not know, Father, how it is, but I know that I should not live long, if I did not breathe. But *how* breathing keeps me alive, I cannot tell. I wish you would explain it. Do we live upon air?

Father.—Yes; we live upon air, as much as we live upon food. And the demand for air is much more constant and urgent, than for food. We can live *weeks* without eating, but we cannot live *minutes* without breathing.

Frank.—I know it is so, Father; but how can the lungs, drawing in, and blowing out air, like a great bellows, support life?

Father.—The lungs are one of the most important organs of the body, and there are no others, if we except the heart and brain, upon the function of which, life so immediately depends. Do you recollect that St. Paul, in his memorable speech to the Athenians on Mar's hill, says, "He giveth us life and *breath*, and all things!" Here, St. Paul speaks of *breath* as one of the *peculiar gifts* of God. I will now explain the manner in which air, taken into the lungs, sustains life.

The air we breathe, called atmospheric air, is composed of two gases, called oxygen and nitrogen gas, in the proportions of about twenty-one parts oxygen, and seventy-nine nitrogen, with the addition of a small percentage of carbonic acid and vapor. Oxygen is the vital principle of the air, and supports all animal and vegetable life. The oxygen is taken up by the blood in the lungs, and conveyed through all parts of the body.

Frank.—Father, how does the oxygen get into the blood?

Father.—You recollect what I told you relative to the minute air-cells of the lungs,—that the inner surface of these cells, if spread out, would be thirty times greater than the surface of the body, or would be larger than the floor of a room, twenty-five feet square.

Frank.—Yes, I recollect that, but what has that to do with the oxygen of the air getting into the blood?

Father.—It has a great deal to do with it, as I was about to explain.

Oxygen has a greater affinity for, or is more disposed to unite with blood, than to remain united with nitrogen; and as the blood passes through the minute blood-vessels of the air-cells, which are very thin membranes, the oxygen separates from the nitrogen, passes through this membrane, and is absorbed by the blood; 'and this large extent of surface of the lungs, or air-cells, is to bring and keep a large amount of air

How breathing sustains life.

in contact with the blood. The blood, you recollect, loses its bright vermilion color, while circulating in the capillaries, and in the veins it presents a dark purple color, or is nearly black.

The cause of this dark color of the blood is as follows:—A poisonous agent, called carbonic acid, is formed in the system, and is taken up by the blood, while passing through the capillaries, and this renders the blood unfit for further use until the carbonic acid is, in some way removed; and this is accomplished in the following manner.

The dark venous blood, containing the carbonic acid, is carried to the heart by the veins, and, by the action of the heart, is conveyed to the lungs; and the carbonic acid, having a greater affinity for, and being more disposed to unite with atmospheric air, than to remain with the blood, passes out through the sides of the air-cells and mingles with the air in the lungs, and is thrown off from them with the breath.

These two important objects are accomplished by every breath of air taken into, and expelled from the lungs,—the vital principle of the air—the oxygen, is absorbed, or taken up by the blood, and the poisonous carbonic acid is thrown off. And this process is going on during every moment of life, asleep or awake, —absorbing oxygen with every breath inhaled, and throwing off carbon with every breath exhaled.

In this way, the dark colored, impure blood, is

relieved of its impurities, and by absorbing the oxygen of the air, again assumes the bright scarlet color, and is endowed anew with the principle essential to life.

Fred.—How wonderful are the ways and works of God! But I wish to ask, how much air is taken into the lungs at each breath, and how much oxygen is taken up by the blood?

Father.—A common sized man inhales about forty cubic, or solid inches of air, at each inspiration, and the blood absorbs two cubic inches of oxygen, which is thirty-six cubic inches a minute, or twenty-one hundred and sixty per hour, or thirty cubic feet per day.

Fred.—Does the blood absorb all the oxygen of the air taken into the lungs?

Father.—No, only about one-fourth of it is taken up by the blood at the first inspiration. But if we inhale the same air several successive times, additional portions of the oxygen will be absorbed each time.

Frank.—And how much carbonic acid is thrown off?

Father.—There is a less quantity of carbon exhaled, or thrown off, than oxygen inhaled, there being about twenty cubic inches per minute, or over sixteen cubic feet per day.

Mary.—Father, how does the poisonous carbon get into the blood?

The blood, how purified.

Father.—It is generated or produced by the changes that take place during the process of nutrition. And had there not been some way provided by the Creator, by which it could be thrown off, it would cause death in a few moments. In proof of this, I need only cite the fact, that if one ceases to breathe, or "holds the breath," for only a few seconds, the carbon retained in the blood, which should have been thrown off with the breath, causes the face to turn purple or black.

I will now repeat, in a few words, that you may the better understand it, the substance of what I have said in regard to the blood.

In the arteries, it is of a bright scarlet color, containing all the elements of nutrition obtained both from the food and from the air while passing through the lungs. While circulating through the capillaries, the nutriment is absorbed, or taken up by the tissues, or parts through which it flows, and at the same time, the blood is filled with carbonic acid,—a poisonous gas, which renders the blood unfit for further use until it is relieved of the carbon.

It is then returned by the veins, to the heart and lungs, where the carbon is taken up by, or unites with, the air in the lungs, and is thrown off with the breath. And this process is going on during every moment of life. The blood is absorbing, with every breath, the vital principle from the air in the lungs, and carrying it to, and depositing it, in every part of the body, giving life and vigor to the system; and is also, with every breath, throwing off the poison, which, if retained in the blood, would instantly cause death. We might in truth say, that with every breath we inhale life, and exhale death.

And what kindness it was in our Maker, so to form the lungs, and all the parts concerned, that this vital process goes on without a moment's cessation, by day and by night, without a thought, or a moment's care of ours.

Fred.—How stupid and thoughtless I am,—always to have been breathing, and never before to have known, or even to have thought, of the wonderful facts you have told us!

But I wish to ask what becomes of the carbon that is constantly thrown off by the lungs? I should think the air would become poisonous in the course of time, and unfit for use.

Father.—It certainly would, had not God, in the arrangements of his works, appointed a beneficial use for what otherwise would have been destructive in its effects.

Carbon, that is poisonous to man and beast, is the *food* upon which *vegetables* live. It is the nutriment of trees and plants, as oxygen is of man and beast. And while all vegetables absorb carbon, which is exhaled or thrown off by animals, *they*, in turn, exhale

Oxygen, support of animal life.

oxygen for the support of animal life; and thus the balance is preserved between the amount of oxygen required for man, and carbon for vegetables, and the air is thus kept in a condition to sustain both animal and vegetable life.

There is another fact to which I wish to call your attention,—a fact, of which, perhaps, you have never thought.

To sustain life, we must breathe every moment; but we need food but two or three times in twenty-four hours. Now think of the goodness manifested in providing for these wants.

The air, without which we could scarcely live one minute, surrounds, envelops, and presses upon us, as the water does upon the fish in the ocean. We could not get away from it if we would. It is ever present with us, and always ready for use. It surrounds the whole earth, and rises forty or fifty miles above its surface.

But food, which we need to use but seldom, is provided in small quantities, and we keep it laid by to be used when needed. Were a man obliged to travel where he could get no food, he could easily take with him a supply for two or three days. But if he were obliged to take a supply of air for two days, it would require more than sixty hogsheads—a rather cumbersome load.

Moral reflection.

Let us now close this social interview, impressed with the wonderful provision our heavenly Father has made for our physical, or bodily wants, and for the continuance of life, and the enjoyment of health.

QUESTIONS ON CONVERSATION X.

PAGE 140.—What is said on this page of the process of digestion?

PAGE 141.—What must now take place? In what does it consist? To what must the blood be conveyed? What does this require? What is requisite for the circulation of the blood? Describe the heart; the arteries; the capillaries; the veins.

PAGE 142.—Where does the blood vary in color and quality? What is its color in the arteries? Where does it lose its scarlet color? What is its color in the veins? To understand the cause of these changes in the color and quality of the blood, what must you know? What are the organs of respiration? Where are they situated? What is their form? Into what are they divided? Into how many lobes is each lung divided? By what are they separated from each other? What is the color of the lungs?

PAGE 143.—In what is each lung enclosed? What is inflammation of this membrane called? Of what are the lungs formed?

PAGE 144.—By what are the lungs retained in their place? Describe the trachea. What is said of the number of air cells? What of the inner surface of these cells? Of what is the trachea made up?

PAGE 145.—By what is it kept open? Why is the windpipe made of bony rings?

PAGE 146.—Describe the bronchial tubes? In what do they end? What takes place in the act of breathing?

PAGE 147.—What benefit is there in drawing air into the lungs and throwing it out again? What is said about living upon air? What is said of the lungs? Of what is the air we breathe composed?

PAGE 148.—In what proportions? What is oxygen? By what is it taken up? Where is it conveyed? How does the oxygen get into the blood? What purpose is subserved by the large extent of surface of the air cells?

PAGE 149.—What causes the dark color of the blood in the veins? What is said of the carbonic acid? What effect has the carbon upon the blood? Describe the process by which the carbonic acid is removed from the blood. What two important objects are accomplished by every breath? What is said of this process?

PAGE 150.—In what way is the blood endowed anew with the principle of life? How much air is taken into the lungs, and how much oxygen is taken up by the blood at each breath? How much carbonic acid is thrown off, or exhaled, at each breath?

PAGE 151.—How does carbonic acid get into the blood? What would have been the result, had no way been provided for throwing it off? What proof is cited? Repeat what has been said in regard to the blood.

PAGE 152.—Do you discover in the formation and function of the lungs, any evidence of the kindness and goodness in the Creator? Why does not the air, in the course of time, become poisonous by an excess of carbon, and unfit for use? What other interesting fact is named on page 153?

PAGE 154.—With what should we be impressed?

CONVERSATION XI.

THE CIRCULATION OF THE BLOOD.

THE HEART, A DESCRIPTION OF—THE HEART A DOUBLE ORGAN—THE AURICLES AND VENTRICLES OF THE HEART—THE FUNCTION OF THE HEART—ITS MODE OF ACTION—THE VALVES OF THE HEART—THE ARTERIES COMMUNICATE WITH EACH OTHER—THE VELOCITY OF THE BLOOD IN THE ARTERIES—DESCRIPTION OF THE CAPILLARIES—CAPILLARY CIRCULATION—DESCRIPTION OF THE VEINS, THEIR FUNCTION.

Father.—Having given you, at our last meeting, a brief description of the structure and function of the lungs, and the changes produced in the blood while passing through them, by absorbing the vital and life-giving qualities of the air;—and imparting to the air in the lungs, and throwing off with the breath, its poisonous and deadly qualities,—I shall now resume a description of the *circulation of the blood.*

The blood may now be regarded as a nutritive fluid, containing all the necessary principles or ingredients for the formation and nourishment of every part of the body.

Fred.—You have repeatedly spoken of the circulation of the blood, and of the vessels in which it flows,

The heart a hollow organ.

but you have never told us what *causes* the blood to circulate.

Father.—The organ provided expressly for that purpose, is the HEART, which is one of the most wonderful organs of the body, whether we consider its mechanical structure, or the importance of its function. It is placed in the left side of the chest, between the right and left lung. It is shaped somewhat like a cone with the small end down.

Fig. 31.

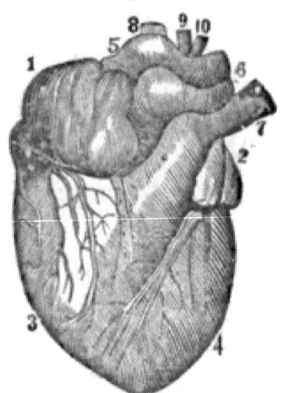

THE HEART.

Fig. 31. 1, is the right auricle; 2, the left auricle; 3, the right ventricle; 4, the left ventricle; 5, 6, 7, 8, 9, 10, the vessels which bring blood to, and carry it away from the heart.

The heart is composed principally of muscular fibres, which run in different directions, some longitudinally, or lengthwise, but most of them around, or in a spiral direction.

The heart is a hollow organ, having four cavities, or rather it is a double organ having two sides, the

Auricles and ventricles.

right and left, each side having two cavities, called the *Au'-ri-cle* and the *Ven'-tri-cle*, which communicate with each other. The auricles are connected with the veins, and the ventricles with the arteries. The right auricle is connected with, and receives the blood from, the largest vein in the body, formed by the union of the ascending and descending *Ve'-na Ca'-va*, which will be more fully explained hereafter.

The *right* ventricle is connected with the *Pul'-mon-a-ry* artery, which carries the blood from the heart to the lungs. The *left* auricle is connected with the pulmonary veins, which return the blood from the lungs to the heart, and the left ventricle is connected with the largest artery in the body, called the *A-ort'-a*, which, with its numerous branches, distributes the blood to all parts of the body.

Fred.—But, Father, you do not tell us how the heart causes the blood to circulate. You say it is hollow, and has four cavities, but that explains nothing, for the arteries and veins are hollow too. Hollow organs may allow the blood to flow through them, but what is it that forces the blood along?

Father.—I have told you, my son, that it was the heart, and that the heart is a muscular organ, and that most of the muscular fibres run around it. Now what have I told you in regard to the action of muscles? in what way do they perform their action?

Fred.—By contracting and relaxing.

Father.—Well, if the muscles, which run around the heart contract, what will be the effect?

Fred.—O, I see now, Father. The muscles contracting, draw the heart together and make the cavities smaller, and if they are filled with blood, it must force it out. Is that right?

Father.—That is correct, and I am pleased to see that you understand it.

The alternate contraction and relaxation of the heart, —what is called beating, or pulsation of the heart, takes place about seventy-five or eighty times a minute, forcing from each ventricle about two ounces of blood at each contraction or pulsation.

I will now explain the action of the heart, as I perceive you are now able to understand it.

The blood is received from the veins into the right auricle, which, by contracting, forces it into the right ventricle. The ventricle then contracts, and forces the blood through the pulmonary artery into the lungs.

Frank.—Does the auricle remain contracted until the ventricle contracts?

Father.—No, the auricle relaxes when the ventricle contracts.

Frank.—Why then does not the blood flow back into the auricle, instead of going to the lungs?

Father.—Because God has provided a valve between the auricle and ventricle, called the *tri-cus'-pid*, which allows the blood to flow from the auricle to the

RENDERED ATTRACTIVE.   159

Circulation illustrated.

ventricle, but when the ventricle contracts, the valve closes and prevents the blood from flowing back into

Fig. 32.

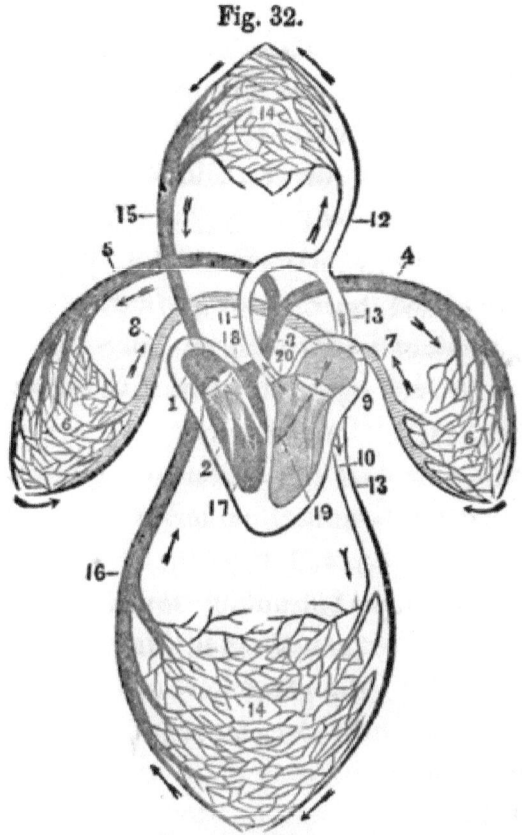

Fig. 32. An ideal illustration of the whole circulation. In explaining this figure, I shall repeat some things I have already said in explaining the circulation, hoping thereby to make it plain.—From the right ventricle of the heart, 2, the dark blood is thrown into the pulmonary artery, 3, and its branches, 4, 4, carry it to both lungs. In the capillary vessels, 6, 6, the blood comes in contact with the air, and becomes red and vitalized. Thence it is returned to the left auricle of the heart, 9, by the veins, 7, 8. Thence it passes into the left ventricle, 10. A forcible contraction of this sends it forward into the aorta, 11. Its, 12, 13, 13, distribute it to all parts of the body. The arteries terminate in the capillaries, 14, 14. Here the blood loses its redness, and goes back to the right auricle, 1, by two great veins, 15, 16, called vena cava, one from the upper, and the other from the lower extremities. The tricuspid valves, 17, prevent the re-flow of the blood from the right ventricle to the right auricle.

the auricle; and as it cannot flow back, it is forced into the pulmonary artery, and the ventricle then relaxes.

Frank.—I should think, that when the ventricle relaxes, the blood would flow back into it from the pulmonary artery.

Father.—So it would, had there been nothing to prevent. But the Creator has placed another valve,—the *sem-i-lu'-nar*, at the mouth of the pulmonary artery, which allows the blood to flow from the ventricle to the artery, but closes when the ventricle relaxes, and thus prevents the blood from flowing back, and conseqently it is driven forward into the lungs, where it unloads the poisonous carbonic acid, and absorbs from the air, the life-giving oxygen, as I have before explained.

Thus far, the heart has only received the blood from the veins and conveyed it to the lungs,—and this is all the *right* side of the heart has to do. But the blood being now fitted to nourish the system, must be circulated through every part of the body, and this important function is assigned to the *left* side of the heart.

The blood is now collected from the lungs by the pulmonary veins, and returned to the *left* auricle, the contraction of which throws it into the left ventricle. The left ventricle now contracts, closing the *mi'-tral* valve placed between it and the auricle, and forces the blood into the aorta—the main artery of the body, at the mouth of which is another semilunar valve, which prevents the blood from flowing back from the aorta to the ventricle.

This artery divides and subdivides, like the branches of a tree, into extremely small vessels, which distribute the blood to every part of the system, as I have before said.

Thus you see the blood passes twice through the heart while making one revolution through the body.

1st. It passes from the right side of the heart to the lungs, to be vitalized by the air.

2d. It then goes, through the pulmonary veins, to the *left* side of the heart, the contraction of which propels it through thousands of tubes situated in every part of the body.

The passage of the blood from the right side of the heart, through the lungs, and its return to the left side, is called the *pulmonic*, or *lesser circulation*. Its passage from the left side through all portions of the body, and its return to the right side, is the *systemic*, or *greater circulation*.

Frank.—Father, it seems to me, that the left side of the heart has much more labor to do than the right side;—for it must require a much greater force to propel the blood through the entire body, than simply through the lungs.

Father.—You are correct, my son. The force re-

Description of the arteries.

quired to be exerted by the left side of the heart is *far greater* than that of the right. And, in accordance with this fact, we find the walls of the left ventricle, much thicker, and therefore capable of exerting a much greater force, than those of the right. But I must now give you a description of the arteries.

The ARTERIES are tubes composed of three coats. The external or outside coat is firm and strong. The middle, consisting of fibres, is elastic, and thicker than the external one. The inside coat is a thin membrane which lines the interior of the artery, and gives it a smooth, polished surface, so that the blood may flow through it without friction.

The arteries communicate with each other freely,— that is, branches run off from one to another, forming a connection between the two.

Frank.—But why do the arteries run together and form communications with each other? You told us that they kept dividing like the branches of a tree.

Father.—While we are pursuing the various occupations of life, the arteries are liable to be compressed, so as to prevent a free flow of blood through them.

This would be injurious, if not fatal, and the Creator, therefore, has wisely and kindly formed this connection of the arteries with each other, so that the circulation can still go on.—There is another reason why the arteries should communicate with each other.

Danger of wounding an artery.

The *Surgeon*, in case of injury, is sometimes obliged to tie up an artery so as permanently to prevent the flow of blood through it.

What then would be the condition of the part, had there not been a way provided by which the blood, as it accumulates in the wounded artery, could flow off into others? It cannot return to the heart and vacate the artery, for the heart, at every pulsation, is throwing the blood into it with great force. It must, therefore, remain stagnant, and would soon putrify and destroy the life of the part, if not of the whole body.

There is another interesting fact. To injure an artery by a cut, bruise, or in any other way, is far more dangerous than to wound a vein.

The arteries, therefore, as a means of safety, are laid deep in the flesh, near the bones, and in some places the bones are grooved out to form a passage for them, so that if the part should be cut to the bone, the artery would remain uninjured.

Fred.—These facts certainly show the kind care of our heavenly Father, in guarding our bodies from accident and danger, and providing for our safety. But I wish to ask, why there is more danger in wounding an artery than a vein?

Father.—For two reasons:—1st. By the action of the heart, the blood is pressed into the arteries with great force, as I shall soon explain, and, therefore, if one of them is cut or wounded, the blood rushes out in an impetuous stream, and unless immediately and permanently checked, which is not an easy thing always to do, life is in imminent danger;—whereas, in the veins, the force employed to circulate the blood, is comparatively small, and, therefore, but little danger results from cutting or wounding one of them.

2d. The blood in the arteries contains the life-giving qualities upon which the body lives; whereas the blood in the veins is charged with a poisonous gas which renders it unfit to nourish the body; therefore the loss of a small quantity of arterial blood is a greater injury than the loss of a large quantity of venous blood.

Frank.—You say, Father, that the blood is pressed into the arteries with great force; will you tell us how fast the blood flows in the arteries?

Father.—In the *large* arteries it flows at the rate of sixty feet per minute, or one foot per second, each pulsation or beat of the heart throwing the blood nearly a foot.

Frank.—Father, you alarm me! Is the blood rushing through our bodies at such a fearful rate, and yet we feel nothing of it?

Father.—This is the rate at which the blood flows *when our bodies are in a quiet state.* When excited by violent exercise, as in running, lifting, etc., the velocity or rate of speed is greatly increased, amounting, often, doubtless, to more than a hundred feet per minute. But the rate of speed is not so great in the capillaries and veins, as I shall hereafter show.

Fred.—How can *exercise* cause the blood to flow with greater speed? I thought you said it was the heart that caused the blood to circulate through the body.

Father.—You have proposed a question upon a very interesting subject, my son, and I will answer it when giving a description of the veins, which I shall do before we close this conversation; but I must first give a brief description of the Capillaries.

The CAPILLARIES are minute tubes, situated between the small arteries and veins, and form a communication between the two. They form a perfect network throughout every part of the body, and are so very small that they can be seen only by the use of the microscope,—their diameter being only one three-thousandth of an inch,—or, in other words, three thousand capillary tubes, laid side by side, would extend but one inch, and these are the extremely small blood vessels to which I referred, when speaking of the lacteals that absorb the chyle.

The capillaries are very numerous, there being many millions, and are situated in every organ and fibre of the body. The finest needle, introduced beneath the skin, is sure to wound several of these minute vessels, from which the blood instantly starts.

Nutriment obtained only in the capillary circulation.

One remarkable fact in regard to the capillary circulation is, that it has no particular and definite direction, but flows in every possible and conceivable direction, thus bringing the blood, while in the capillaries, in contact with every particle of the body.

Frank.—But why is it, Father, that the circulation of the blood in the capillaries, is so different from that of the arteries and veins?

Father.—It is for the reason that the body receives nutriment from the blood *only* in the capillary circulation. The arteries and veins are simply tubes or pipes, in which the blood flows from one part of the body to another; the same as the water flows from the Cochituate lake to this city, through the large pipes laid for that purpose.

The body can no more obtain nutriment from the blood while circulating in the arteries and veins, than the citizens of Boston can obtain benefit from the Cochituate water while flowing from the lake to the city in the water pipes.

It is during the capillary circulation alone, that the system obtains nourishment from the blood; and, therefore, it is very slow,—being only two inches per minute, thus affording time for the absorption of nutriment by the tissues of the body.

The *color* of the blood is seen only in the capillaries. If the hands, or any part of the body, be immersed, for a short time, in warm water, the surface becomes red, in consequence of a rush of blood into the capillaries.

A glow of health often renders the countenance so florid and blooming, that, if the attempt were ever made by ladies, to excel nature by art, we should sometimes be at a loss to decide whether the tints of the cheeks were natural or artificial.

The capillary circulation is affected greatly by *mental emotion*. Thus a sense of shame causes a rush of blood to the surface, producing what is called, a *blush*. The sensation of terror, or fright, drives the blood from the surface, causing a pallid countenance.

The VEINS, as I have already informed you, are the vessels which return the blood to the heart, after it has been circulated through the arteries and capillaries, to all parts of the body.

They originate in, or are connected with, the capillaries, by minute branches, which, running together, form tubes larger and larger, until they all unite in two large veins, called the ascending and descending *Ve'-na Ca'-va*, the descending returning the blood from the upper, and the ascending from the lower extremities of the body, and uniting and forming one vein just as it enters the right auricle of the heart.

The veins are provided with valves, which allow the blood to flow freely from the capillaries to the heart, but prevent its returning from the heart towards

Veins supplied with valves.

the capillaries, and in this respect, are unlike the arteries, which have no valves.

Fred.—I do not see, Father, what good valves can do in the veins, as there is nothing to force the blood back from the heart, and yet, if there *are* valves, I know they are placed there for an important purpose. Will you please tell us what it is?

Father.—I will explain it, and in doing so, I shall answer the question you proposed, not long since, " how exercise increases the velocity of the circulation of the blood."

The veins are situated among the voluntary muscles, which, a great part of the time, are in a state of alternate contraction and relaxation. At every contraction, these muscles become rigid, or hard and swollen, and consequently press upon the veins which run between them. This pressure upon the veins, forces the blood out of that part upon which the pressure is made, and as the valves prevent its returning back, it must, of course, be forced onward towards the heart. When the muscle relaxes and relieves the vein from the pressure, it again fills with blood, and the muscle again contracts, and forces the blood forward as before. And the more frequent the contraction of the muscles, or, in other words, the more violent the movements of the body, the more rapidly will the blood be forced along the veins, to the heart.

Frank.—I see now, Father, how bodily exercise

quickens the flow of blood in the veins; but I cannot see how it can increase the action of the heart, so as to cause it to beat more rapidly, as I have noticed it always does.

Father.—If the pressure upon the veins by the action of the muscles which I have described, causes the blood to flow more rapidly through the veins, it will, of course, be forced more abundantly into the heart, and the heart must necessarily increase its exertion to keep itself clear,—as the crew of a leaking, sinking ship, increase the action of the pumps when the water rushes in so as to endanger their lives;—the increased quantity of blood acting as a stimulus to the heart.

Frank.—I understand now why the heart beats more rapidly, when we take exercise. But it is a mystery to me—and it is what I have thought of a great deal,—why exercise should cause us to *breathe* more frequently. Because breathing, you know, has nothing to do with the circulation of the blood,—it is only taking air into the lungs, and throwing it out again.

Father.—That is a very interesting inquiry, my son. But, as it seems to me, that I have given you sufficient information, to enable you, by close thinking, to solve the problem yourself, and as our conversation has already been too long, I will leave the question for you to think upon till our next meeting, hoping that you will then be able to answer it yourself.

QUESTIONS ON CONVERSATION XI.

PAGE 155.—How may the blood be regarded?

PAGE 156.—What organ causes the blood to circulate? What is said of the heart? Where is it placed? What is its shape? Of what is it composed? In what direction do the fibres run? Describe the cavities of the heart.

PAGE 157.—What are the cavities called? With what are the auricles connected? The ventricles with what? From what does the right auricle receive the blood? By the union of what is this vein formed? With what is the right ventricle connected? With what is the left auricle connected? With what is the left ventricle connected? What is the function of the aorta and its branches?

PAGE 158.—Explain the action of the heart in forcing the blood through the system. What is the alternate contraction and relaxation of the heart called? How often does it take place? How much blood is forced from the ventricles at each pulsation? Give a minute description of the action of the various parts of the heart, with the names and functions of its valves. Where is the tricuspid valve placed? Where is the semilunar placed?

PAGE 160.—Where is the blood driven? Only what has the heart thus far done? What must now be done with the blood? To what is this function assigned? By what is the blood collected from the lungs? What is done by the left auricle and ventricle?

PAGE 161.—Where is another semilunar valve placed? For what purpose? Into what does this artery—the aorta, divide? What is done by these small vessels? Does the blood pass through the heart more frequently than through the body? Repeat what is said in regard to it. What is the pulmonia, or lesser circulation? What is the systemic or greater circulation.

PAGE 162.—What is said of the force of the left side of the heart? What of the walls of the left ventricle?

PAGE 163.—What are arteries, and how composed? Describe the three coats. Arteries communicate with what? Why do arteries communicate with each other? What is the surgeon obliged to do? What would have been the result, had not this communication of the arteries with each other been made? In what way is the safety of the arteries provided for? What do these facts show?

PAGE 164.—What two reasons are given why there is more danger in wounding an artery than a vein? At what rate does the blood flow through the large arteries? When and how is this rate or speed increased? What is said of the rate of speed in the capillaries and veins?

ANATOMY AND PHYSIOLOGY.—CONVERSATION XI.

PAGE 165.—Describe the capillaries. What do they form? What is their diameter? What is said of the number of capillaries? Where are they situated? What would wound several of them?

PAGE 166.—What remarkable fact is there in regard to the capillary circulation? What reason is assigned for this? What is said of the arteries and veins? At what rate does the blood flow in the capillaries? Why is the circulation so slow? In what is the color of the blood seen? What is the proof?

PAGE 167.—What is said of a glow of health? By what is the capillary circulation affected? Give examples. What is the function of the veins? Describe their origin and formation. With what are they provided? For what purpose?

PAGE 168.—Where are the veins situated? What is the effect of the pressure of the muscles upon the veins? What forces the blood more rapidly along the veins?

PAGE 169.—How does exercise increase the action of the heart? Why does exercise cause us to breathe more frequently?

CONVERSATION XII.

THE CIRCULATION OF THE BLOOD—CONTINUED.

The force exerted by the heart—The heart never becomes fatigued—The quantity of blood that passes through the heart—The heart-case—Quantity of blood in man—Animal heat, theories as to how it is produced—Life depends on a uniform standard of heat—Animal heat the same in all countries.

Father.—Well, my dear children, I hope you are prepared to answer the question proposed by Frank at the close of our last conversation:—"why we breathe more frequently when taking exercise than at other times."

Frank.—We have been thinking upon it, Father, and talking it over; and I think we can give the right answer,—but we are not certain until you answer a question which we wish to propose.

Father.—Well, Frank, what is the question?

Frank.—We wish to know if the blood has always the same amount of oxygen and carbon in it.

Father.—I do not understand you. Please make out your meaning.

Frank.—I mean—that—

Why exercise causes frequent breathing.

Fred.—Please let me state the question, Father. I think I can make you understand it.

We want to know, if the blood, as it comes from the heart to the lungs, has the same amount of carbon in it, whether it comes slowly or fast; and, also, if it absorbs the same quantity of oxygen,—that is, does a given amount of blood absorb the same amount of oxygen, whether it passes through the lungs rapidly or slowly?

Father.—I understand you now. Frederick has stated it very clearly, and I am beginning to think, from the question, that you have the correct answer.

The blood *does* contain the same amount of carbon, and absorbs the same quantity of oxygen, whether it flows rapidly or slowly.

Frank.—Then it is plain, that if the blood flows more rapidly through the lungs during active exercise, the breathing also, must be more frequent, that the blood may be supplied with oxygen, and, also, that the carbon may be conveyed from the lungs.

Father.—I am happy to assure you that your answer is correct. The velocity, or rapidity of the circulation of the blood, and the frequency of breathing, are always in perfect accordance, and it is so for the reasons that you have stated.

I now wish to call your attention to some interesting facts in regard to the action of the heart, which I have purposely deferred until I had made you ac-

Organs of the chest and abdomen.

Fig. 33.

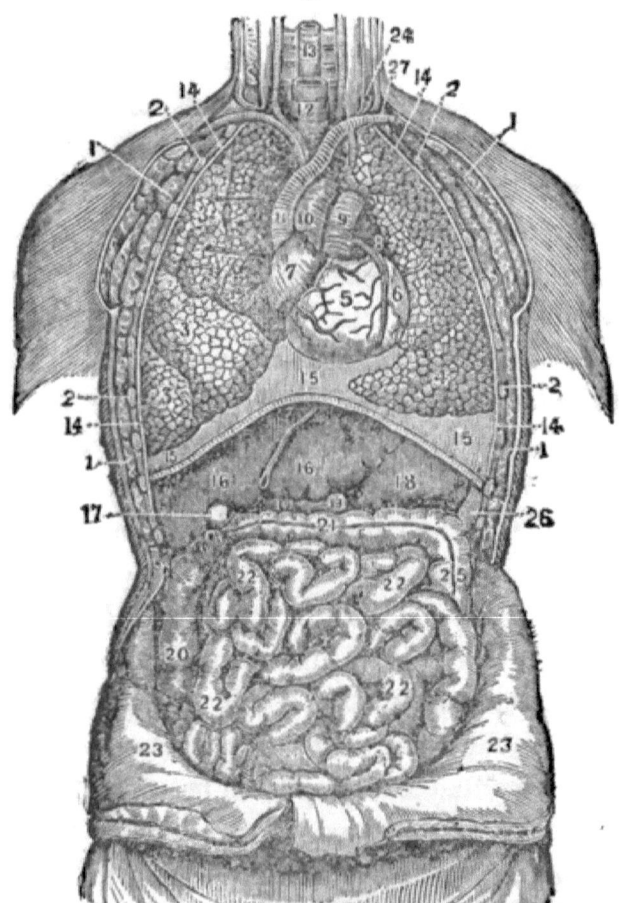

A FRONT VIEW OF MANY OF THE ORGANS BOTH OF THE CHEST AND ABDOMEN.

Fig. 33. 1, 1, 1, 1, are the muscles of the chest; 2, 2, 2, 2, the ribs; 3, 3, 3, the upper, middle, and lower lobes of the right lung; 4, 4, the lobes of the left lung; 5, the right ventricle of the heart; 6, the left ventricle; 7, the right auricle of the heart; 8, the left auricle; 9, the pulmonary artery; 10, the aorta; 11, the descending vena cava; 12, the windpipe; 13, the œsophagus; 14, 14, 14, 14, the pleura; 15, 15, 15, the diaphragm; 16, 16, the right and left lobes of the liver; 17, the gall bladder; 18, the stomach; 26, the spleen; 19, 19, the duodenum; 20, the ascending colon; 21, the transverse colon; 25, the descending colon; 22, 22, 22, 22, the small intestine; 23, 23, the walls of the abdomen turned down; 24, the thoracic duct, opening into the left subclavian vein (27).

quainted with the facts contained in our last conversation.

I have already informed you that the alternate contraction and relaxation of the muscles of the heart, is the principal agent or cause of the circulation of the blood through the body. There has been a variety of opinions among the learned, as to the amount of force required to be exerted by the heart, to propel the blood through so many millions of minute vessels. The force exerted has been variously estimated, as equal to from sixty to one thousand pounds weight; but some of the most learned place it at *four hundred pounds*. We will, therefore, suppose *that* to be correct.

Now, just think of the following facts.

Here is a muscular organ, exerting a force,—not once an hour, nor once a minute, but *seventy-five or eighty times* a minute, equal to *four hundred pounds weight*,—and that organ *never becomes fatigued*, although its rapid action may be continued for *seventy-five or a hundred years*, Nor could it stop one moment to rest, if it were fatigued, for should it stop, that moment we should die.

I know of nothing in nature that will explain the fact, that a muscular organ is exerting a force equal to four hundred pounds weight, seventy-five times a minute, for fifty or one hundred years, and *never becomes fatigued!*

How the heart wears.

Fred.—Why, Father! how can that be?

Father.—I do not know, Frederick. The only answer I can give is, that God saw that it was necessary—for we could not live without it—and, therefore, made the heart with capacities different from all other muscular organs of the body; for, let one attempt to keep, even a ten pound weight in motion with the hand, and how soon the arm will become fatigued.

Again, one would suppose that such a force exerted by an organ within the system, would send a thrill of pain, at each pulsation, through every fibre of the body, and would shake and convulse the body with awful tremors. But what is the fact? Why, so even, and uniform, and smooth, is the action of the heart, that not one person in a hundred, perhaps, is reminded once a month, that he has a heart.

Again, think how the heart wears. Were it made of steel, it would not last a year; the valves, moved with such a force, would wear out. But the heart may continue in motion for a hundred years, and still be perfect in every part.

Then, think of the quantity of blood which passes through the heart. I have already stated, that more than two ounces of blood, is expelled from the heart, into the arteries, at each pulsation. Consequently, not less than nine thousand six hundred ounces are thrown into the arteries and propelled through the body in one hour, which would amount to one thousand four hundred pounds per day, and in seventy-five years, would form a pond of blood more than a quarter of a mile square, and five feet deep. If you will place your ear upon the left breast of a person, over the heart, you will hear a sound like the *rushing* of a *mighty cataract.*

Mary.—Why, Father, you frighten me!

Father.—These facts do seem somewhat frightful, Mary, because the least derangement in any of this wonderful and complicated action of the heart, would cause immediate death. The numerous cases of persons dropping down dead, is proof of the truth of this statement.

Fred.—I should think, Father, that the heart would be chafed and irritated, by rubbing against the surrounding parts, as it is all the time in motion.

Father.—Such, no doubt, would be the case, had not its safety been specially provided for. The Creator, knowing the value of the heart, has protected it from harm, by surrounding it with a sac, called the *Per'-i-car'-di-um*, or heart-case, the inner surface of which, secretes a slippery fluid, that lubricates the outside, or exterior of the heart, and thereby prevents friction between it and the heart-case. In a healthy condition, there is about a teaspoonful of this fluid in the heart-case.

Now, my dear children, can we fully appreciate the goodness of the Creator, in so forming the heart, that

Facts about the blood.

it performs all the wonderful and important functions which I have described, *entirely independent of our effort or care?*

How long do you think we should have lived, had any part of this process been entrusted to us?

Fred.—I do not think, Father, that I have ever appreciated the goodness of God, in regard to these truly interesting and wonderful facts. But I hope to do better in the future, and not live so thoughtless of the goodness that keeps me alive every moment.

Father.—I am glad to hear you say so, for if these facts will not impress us with a sense of God's goodness, and our duty to love and obey him, I know not what will.

I wish now to make you acquainted with some interesting facts in regard to the blood.

I have briefly explained the manner in which the blood is formed and circulated through the body; by which, you see, that it contains both the nutriment of the food, and the vital principle of the air. It is to the body, what the sap is to the tree—its life.

Fred.—Yes, Father, you have made that very plain; I think I understand it. And how wonderful it is! Certainly, none but God could do it! But you have not told us *how much* blood there is in the body. I should be glad to know. I think there must be a great deal, for if I prick, or injure myself, the blood starts out immediately, and it appears as if the body was full of it.

Quantity of blood.

Father.—The quantity of blood in a medium sized man, is estimated at from twenty-four to thirty pounds, or between three and four gallons, all of which, it is supposed, passes through the system every four minutes.

But what I was about to speak of in particular, is the natural warmth of the blood, or what is called *animal heat,* which consists in the ability of animal bodies—such as man and beast, to maintain nearly a uniform temperature, whether the surrounding atmosphere is cold or hot. You know if you should take a bar of iron, and heat it to two or three hundred degrees, and then place it in air at a temperature of thirty or forty degrees, it would soon cool down to the temperature of the surrounding air, and there it would remain, and would change in temperature only as that of the air changed. But the case is very different with living animals. The temperature of the blood in man, in a healthy state, is about 98 deg. of Farenheit, and it remains the same in summer and winter. But as the temperature of the air is nearly always below that of the human body, and for the greater part of the time, is *far* below, the heat of the body must be continually passing off into the surrounding atmosphere; yet, notwithstanding the body is all the time losing heat, it does not cool down, like the bar of iron, but still maintains its uniform standard of heat. The extremities of the body, it is true, on a very cold day, may become cooled down considerably below the standard temperature; but the internal organs and the blood, remain the same under all ordinary exposure. It is, therefore, evident, that this vital, or animal heat, which is so constant and uniform, must, in some way, be generated in the body.

Now, can you tell me how this heat originates?

Frank.—No, Father! That is what I never thought of before. Do tell us.

Father.—The origin of animal heat has been a fruitful theme of discussion among the learned, and various theories have been proposed to account for it;—but none, as yet, have given a satisfactory explanation.

I will explain to you the most plausible theory,—that which for a time seemed satisfactory to physiologists;—but, that you may understand it, I must first explain the manner in which heat is produced by *combustion*,—that is, by the burning of wood, or coal, or any other substance. It is produced by the *chemical union* of the oxygen of the air with the carbon of the wood or coal.

You recollect I told you that the venous blood, as it comes into the lungs, is filled with carbon;—and it was believed that the oxygen of the air, inhaled in breathing, coming in contact with the carbon of the blood, produced a slow combustion in the lungs, and thereby generated the heat of the blood.

Theories of animal heat.

Frank.—O, Father! that certainly explains it beautifully,—only I should be afraid the lungs would burn up.

Father.—Well, my son, others, more learned than yourself, have thought "that explained" the origin of heat "beautifully," but they, like yourself, were too hasty in their conclusions.

Frank.—That is just the way, Father. Whenever I think I see, either in my own views, or those of others, a correct solution of a problem, you at once pronounce it an error, and show that it is all wrong. But in this case I have the consolation of having good company,—for you say that others, more learned than myself, have thought so.

But, Father, I do not see how you can prove that it is *not* so, for certainly, oxygen and carbon, the union of which you say produces combustion, come together in the lungs, according to your own explanation.

Father.—The theory that heat is generated in the lungs by combustion, is certainly a very beautiful and convenient one, but, like many others equally fine, is destitute of truth; and has been proved erroneous by the following facts :—

1st. It has been found, that the blood in the lungs, is no warmer than in the other organs.

2d. That the oxygen of the air inspired or taken into the lungs, is taken up by what is called the blood globules, and carried into the general circulation, instead of uniting with the carbon of the blood in the lungs.

Another theory, somewhat similar to the one I have named, is, that the blood receives chyle from the stomach, and oxygen from the lungs, and passes with them to the capillaries, where the oxygen unites with the decayed portions of the body, and changes them into carbonic acid, thus producing combustion, the same as when oxygen unites with wood or coal;—the oxygen of the new blood burning up a portion of the carbon in the capillaries, by which *heat is given out*,—the fresh chyle taking the place of the particles consumed. That the carbonic acid produced by this combustion in the capillaries is thrown off by the lungs and skin, and that thus our bodies are kept warm by myriads of little fires in the capillaries.

According to this theory of warming the body, the stomach provides the fuel, and the lungs the oxygen to consume it, and the arteries convey the fuel and fire to the capillaries, where the combustion takes place. The carbonic acid produced, is carried to the lungs by the veins, and to the skin by the pores, and, like smoke from a chimney, pours out of our mouth and nose, and pores of the skin.

Liebig, the great German chemist, believed animal heat to be produced by the combustion of certain elements of the food, while circulating in the blood. But this, as well as all previous theories, is now regarded as erroneous.

Opinions of the learned.

Fred.—I thought, Father, you were going to tell us how animal heat is produced, but you have only told us how it is *not* produced. What use do you wish us to make of such information?

Father.—This is one of the problems, of which it is easier to prove a negative, than an affirmative.

Mary.—What do you mean, Father, by negative and affirmative?

Father.—By a negative, I mean what is *not*, or what does not exist. By an affirmative, I mean what *is*, or what does exist.

Mary.—Then you mean to say, that it is easier to prove what does *not* produce animal heat, than what *does* produce it.

Father.—Yes, my daughter, that is what I mean.

Fred.—Well, Father, are you not going to tell us how animal heat *is* produced?

Father.—It appears to me to be a problem difficult, if not incapable of solution. No theory, perhaps, will ever be proposed to account for it, that will not be open to objections. It may be one of the secret operations of the Creator, which is beyond the scrutiny of man.

Fred.—You have spoken of the opinions of learned men. Do not they give any other explanations than those you have named?

Father.—Yes, some recent writers upon the subject, believe that the production of animal heat depends upon *certain chemical changes* which take place in various organs and tissues of the body. And this is *probably* the correct theory. But until you are better acquainted with the science of chemistry, I cannot explain it to you fully. But whether this theory will prove more satisfactory than its predecessors, or, whether like them, it is destined to be subverted and overthrown by a subsequent theory, is what I cannot predict with certainty.

But whatever be the means by which the heat of the body is produced, one thing in regard to it is certain, that life depends on its being maintained at about its natural standard. If the cold be so intense and long continued as to affect the general temperature of the blood, so that it becomes reduced more than five or six degrees below its natural standard, death is the inevitable result. When a person dies under these circumstances, he gradually becomes insensible and torpid, and the vital functions finally cease.

This natural warmth of the body, is one of the greatest blessings bestowed upon us by our Creator, and yet it is one, which, perhaps, of all others, we enjoy with the least consciousness of its value. As I have already stated, a slight deviation from the natural standard of animal heat, results in immediate death. Now what would have been our condition, had not the Creator, in his goodness, endowed our bodies with power to generate, without any effort or care of ours,

Animal heat everywhere the same.

that degree of heat requisite to sustain life? Life is sufficiently frail with *present* physical endowments. But, had the body been created *without* this power to generate heat, and, had we, consequently, been obliged to rely upon artificial heat to keep us warm, how long may we suppose life would have continued?

In that case, every apartment of our dwellings must be kept at a temperature of not less than ninety-eight degrees. And we could never leave our warm rooms to attend to the business and duties of life, for exposure to the cold air of winter, for a very few moments, would so reduce the heat of the body, that death would immediately follow.

There is another fact in regard to animal heat, in which the goodness of the Creator is clearly manifested. I refer to its *uniformity;*—by which, I mean, that the standard of heat is the same in all persons in all nations of the earth. The temperature of the blood is the same in Europe as in America,—on the burning plains of Africa, as in the frozen regions of Greenland.

That we may see the value and feel the force of this fact, let us suppose that it had been otherwise,—that there was no uniform standard of animal heat, and that in a family living together in the same house, A has a temperature of one hundred degrees, B of eighty degrees, C of sixty degrees, D of forty degrees, and E of twenty degrees. Now suppose these persons are all in a room in which the thermometer shows the heat to be sixty degrees. Will they all be equally comfortable?

Fred.—Why no, Father! While A and B would be freezing, D and E would be dying with heat.

Father.—That is true. And if A and E, being brothers, should attempt to sleep together in the same bed, A would be a *furnace* to E, and E an iceberg to A. But, now, as the Creator has formed us, to persons in health, the same degree of temperature is equally agreeable to all.

Frank.—O, Father, I never should have thought of these things! How grateful we ought to be, that God has made us as we are. But what would be the effect, if the temperature of the blood, should, by any means, be raised *above* a healthy standard?

Father.—It would be no less fatal to life, than when it falls *below*. If the temperature of the blood be raised but a *few degrees above* the natural standard, death will inevitably follow. Cases sometimes occur, in which intense heat of the sun, combined with vigorous bodily exercise, so elevates the temperature of the blood, that sudden death is the result;—and the person is then said to die of " sun-stroke." And such results would be of very frequent occurrence, and life would be in constant jeopardy in tropical climates, had not the Creator, in his goodness, devised means, and endowed our bodies with power, to carry off this

Latent and sensible heat.

excess of heat. The means devised are as wonderful as they are beneficial; and are based upon a well known principle in Natural Philosophy, which I will explain, as you have, as yet, made but little proficiency in that science. I must first remark, that there are two kinds of heat, or caloric, as it is sometimes called,—*free*, or *sensible*, which can be felt, and *latent*, which cannot be felt. If you will keep this fact in mind, you will understand the principle to which I refer, which is, that solids, in passing into fluids, and fluids into vapor, *absorb heat,*—or sensible heat becomes latent or insensible. On the contrary, vapor, in passing into fluids, and fluids into solids, *give out heat*, or latent heat becomes sensible.

For example:—When ice thaws, the heat which melts it, becomes latent, or is absorbed, or taken up by the water which the melting of the ice produces. And if that water is evaporated, the heat that causes the evaporation is absorbed by the vapor, and becomes latent, and consequently insensible.

If you immerse the hand in water, on removing it, there is a sensation of cold, until the hand becomes dry. This sensation of cold is produced by the heat being absorbed from the hand by the water, as it passes into vapor.

Now, my children, do you understand this principle of Natural Philosophy which I have been explaining.

Fred.—Yes, Father, you have explained it so

clearly, that we should be stupid indeed, if we did not understand it. But I never knew before, why my hand, or any other part of my body that was wet, felt so cold when it was drying. I have noticed, that after putting my hands into warm water, they gradually become cold, until they get dry, after which they become warm again.

Father.—Well, I think now you will understand what I wish to say.

This principle that I have been explaining, produces its effect upon the human body in the following manner:—We are so constituted that a slight elevation of temperature of the body, above the natural standard, *produces perspiration.* This perspiration passes off from the skin, in vapor, and in doing so, absorbs the heat which caused the perspiration, and thereby prevents an increase of heat of the blood, much above a healthy standard, as the excess is carried off as fast as it is generated. Thus each pore of the skin, (and there are millions), is constituted a safety-valve to the body. Should the temperature, by any means, be raised above a healthy standard, this *very excess* of heat opens these valves, and the heat is thus let off, and the temperature brought back to a healthy point. Thus, this excess of heat, *proves its own remedy.*

Fred.—How numerous and how wonderful are the ways and means that God has devised to guard our life, and prolong our earthly existence!

QUESTIONS ON CONVERSATION XII.

What important fact is explained on page 171? Name the particulars.

PAGE 173.—What is stated of the opinions of the learned in regard to the force of the heart? What interesting facts are stated, relating to the action of the heart?

PAGE 174.—What might one suppose would be the effect of such a force exerted within the body? What is said of the wear of the heart? What of the quantity of blood that passes through it?

PAGE 175.—What would be the effect of the least derangement of this action of the heart? How is this proved? What prevents chafing of the heart? Describe the heart-case and its function.

PAGE 176.—Do we appreciate the goodness of God in all this? particularly in so forming and endowing the heart that it performs its important functions without our care?

PAGE 177.—What is the quantity of blood in the human body? How often does it pass through the system? What is the natural warmth of the blood called? In what does animal heat consist? What is the temperature of the blood in a healthy state? Is it the same summer and winter?

PAGE 178.—What is therefore evident? What is said of the origin of animal heat? Explain how heat is produced by combustion.

PAGE 179.—Explain the most plausible theory of the origin of animal heat. How has this theory been proved erroneous?

PAGE 180.—Describe another similar theory. What was Liebig's belief?

PAGES 181 AND 182.—The origin of animal heat may be what? What different theory has been proposed by recent writers? What is said as to the correctness of this theory? What one thing in regard to animal heat is certain? Of what is death the inevitable result? What is said of the natural warmth of the body? Give examples.

PAGE 183.—What would result, were we obliged to rely on artificial heat to keep us warm? What is said in regard to the uniformity of animal heat? Give an example of the result, had it been otherwise, or had there been no uniform standard.

PAGE 184.—If the temperature of the blood be raised above a healthy standard, what is the result? What cases sometimes occur? What means has the Creator devised to carry off an excess of heat? Describe the two kinds of heat.

PAGE 185.—What absorbs heat? What gives out heat? Give examples.

PAGE 186.—Explain how this principle produces its effect upon the human body in carrying off heat. This excess of heat proves what? What is wonderful?

CONVERSATION XIII.

THE SECRETORY ORGANS OR GLANDS.

NAMES OF THE PRINCIPAL SECRETORY ORGANS—THE FUNCTION THEY PERFORM—OBJECT AND USE OF SECRETION—DESCRIPTION OF THE PERSPIRATORY GLANDS—NONE OF THE SECRETIONS FOUND IN THE BLOOD—THEORY OF SECRETION—NERVOUS INFLUENCE UPON SECRETION—IMPORTANCE OF THE FUNCTION OF SECRETION—EXCRETORY DUCTS—THEIR USE.

Father.—I have several times spoken of the secretory organs. I will now give you a brief description of them, and the function they perform.

The most important of these organs or glands, are the liver, the kidneys, the salivary glands, the lachrymal, the pancreatic, the perspiratory, the synovial, the glands which secrete the gastric fluid—and several others of less importance—in all not less than twenty.

Each of these glands secretes, or separates from the blood while passing through it, a fluid peculiar to the gland which secretes it, and differing greatly in quality, and in the office it performs, from that of any other.

The *liver* secretes a yellow or dark green fluid,

Fig. 34.

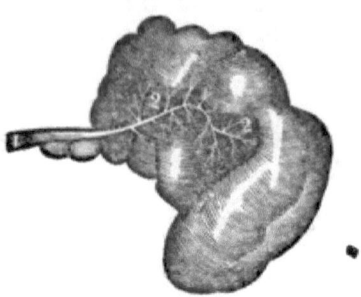

A GLAND.

Fig. 34. 2 2, the small ducts spread through its body, and running together; 1, the large duct, through which the secreted substance is carried away.

called bile or gall, which is a thick, ropy substance, and intensely bitter to the taste.

The *kidneys* secrete a clear, watery, amber-colored fluid, called urine.

The *salivary glands* secrete the saliva—a transparent and tasteless fluid.

The *lachrymal* secrete the tears.

The *pancreatic* secrete the pancreatic juice.

The *perspiratory* secrete the perspiration.

The *synovial* secrete the synovia.

Some of these fluids are designed to fulfil very important purposes in the animal economy, while others seem to subserve the only purpose of removing acrid and irritating substances from the blood, which, if allowed to remain, would soon destroy life;—and others fulfil both these purposes.

Frank.—How can that be, Father? I can understand how some of these fluids can be used for a *good*

Organs that purify the blood.

purpose, as you told us about the gastric fluid dissolving the food in the stomach in the process of digestion. But how can a fluid be both good and bad—I mean, how can it be used for a beneficial purpose, while it is acrid, and if allowed to remain in the blood would destroy life?

Father.—This is one of the many instances in which God accomplishes two or more important objects by one means. The *liver* is one of these organs, the secretion of which—the bile, aids in digestion, as I have previously explained; also, by its acrid qualities, stimulates the bowels to action, thereby keeping them in a regular and healthy condition—and, at the same time, by the secretion of bile, the blood is relieved of a large amount of impurity, which, if retained, would be detrimental to health and life.

Fred.—That is certainly an interesting fact. Will you please tell us which of the secretory organs are designed simply to carry off the impurities of the blood?

Father.—I will name two of the most important,— the *kidneys*, and the *perspiratory glands*.

The kidneys separate from the blood a large amount of acrid material, called urate of soda, urate of potass, urate of ammonia, phosphate of soda, phosphate of potass, phosphate of magnesia, phosphate of lime, etc., held in solution in the secretion of the kidneys, which if retained in the blood, would cause death in a short time. As far as I know, the kidneys were designed solely for the purpose of separating these materials from the blood, and upon their constant and faithful action life depends, and one could live but a few days,—perhaps hours—should their function cease.

The *perspiratory glands* of the skin are small bodies, not more than one four-hundredths of an inch in diameter, and are situated in all parts of the surface of the body, varying in number from five hundred to twenty-seven hundred to the square inch,—the whole number in the body being not less than two million three hundred thousand They secrete, or separate from the blood, a fluid called perspiration, which exists in two forms, called sensible and insensible perspiration,—the sensible appearing occasionally in drops upon the face and other parts of the body, while the insensible is going on continually, but in quantity so minute as to be imperceptible, and, therefore, is called insensible.

Connected with each perspiratory gland, is an excretory duct or tube, whose office, or function, is, to carry off the perspiration secreted by the gland. This tube is spiral in form, and about one-fourth of an inch in length.

Fred.—Father, I would like to ask, how it can be known that perspiration exists, if it is imperceptible?

Father.—The existence of insensible perspiration

Perspiratory gland.

may be demonstrated in various ways. One is, by
Fig. 35.

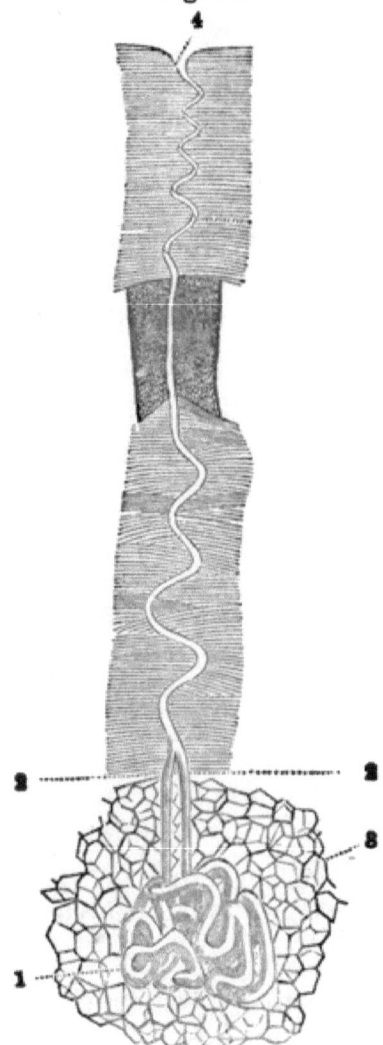

PERSPIRATORY GLAND AND ITS TUBE, GREATLY MAGNIFIED.
Fig. 35. 1, the gland; 2, 2, the two excretory ducts from the gland, which, uniting, form one spiral tube, which opens at 4; 3, are the fat cells.

Mysteries of secretion.

placing the hand near a pane of glass, or a mirror, or a polished surface of steel, and the perspiration will immediately collect on the glass or polished surface.

Another very clear proof of the existence of insensible perspiration is obtained by enclosing the body, or a part of it, say an arm or a leg, in a sack or case of India rubber, through which perspiration cannot pass. In that condition the surface of the limb, and the sack which encloses it, will soon be wet with it.

It is asserted by good authority, that not less than from two to three pounds of insensible perspiration passes from the body of a healthy man in twenty-four hours. In this way, a large amount of impurity is separated from the blood. But as I have explained, in a previous conversation, (See page 186), one important object of perspiration is to regulate the temperature of the body by carrying off excessive heat.

Fred.—You say that all these different fluids are separated or taken from the blood?

Father.—I do.

Fred.—Then I suppose that these fluids or secretions, as you call them, *exist* in the blood, and that each secretory gland separates that particular kind of fluid that belongs to it,—that is, the liver separates the bile from the blood, the kidneys separate their peculiar fluid, the salivary glands the saliva, etc.

Father.—That is a very natural conclusion, my son; but it is a very erroneous one.

The secretions.

Many experiments have been made upon the blood to ascertain the facts upon this subject, but it has never been proved, by the most careful scrutiny of the chemist and physiologist, that the blood contains a particle of bile, gastric juice, saliva, or any other of the secreted fluids.

Fred.—I cannot see how that can be. I should like to know how the liver can secrete, or separate bile from the blood, if there is none in it to be separated,—or the kidneys their secretion, or the salivary glands saliva, if neither of these fluids exist in the blood, from which you say they are taken. Can these fluids be taken from the blood in which they never existed?

Frank.—I see it in the same light that Frederick does. I cannot imagine how these fluids can be taken from the blood if they never existed in it. But I know you have a way of getting over these things and clearing them up,—so that I should not dare dispute or disbelieve any thing you say.

Mary.—It is true, Father, as Frederick and Frank have said, that what you observe about the secretions, seems like a contradiction; and so have statements you have made on other subjects seemed, but these you have finally made all clear to our comprehension; and I have no doubt you can do so in respect to this. Be so good, therefore, as to tell us how the bile and other secretions can be taken from the blood, when they never existed in it.

How the secretions are formed.

Father.—My dear children, this is truly a wonderful fact, and I am not surprised that you are much perplexed with it.

The only explanation that can be given, and which is doubtless the correct one, is the supposition that the blood contains the *elements* of all these different secretions, and each gland is constructed in such a manner, and endowed with the power requisite to separate these elementary principles from the blood, and, by the peculiar action of the glands, to form the elements into the various secretions;—the liver separating from the blood, the elements of which bile is composed,—and so of all the secretions of the body,—each gland being adapted to separate from the blood, the elements which compose its peculiar secretion, and to form them into the fluid secreted.

But what the peculiar structure of each gland is, that causes such extremely different results in their action—one secreting a bitter, another a salt, and another a tasteless fluid, is one of the secret operations of the Creator which is hidden from man.

We can no more explain this fact, than we can explain how sweet apples, can, by ingrafting, be formed from the same sap that produces sour apples.

Fred.—Is the quality of the blood the same from which all these different fluids are taken?

Father.—The quality is the same, and it is all arterial blood, with the exception of the bile, which is secreted from venous blood.

Interesting facts about secretion.

Frank.—This is truly *wonderful*, but I do not see any particular proof of *goodness* in it.

Father.—I am sorry for that, Frank. I think *I* discover very *marked* proof of goodness. I will name one, and by thinking closely, you may discover others.

The saliva, in the mouth, is a perfectly tasteless fluid, which is, perhaps, the only tasteless secretion in the body, while some are extremely acrid, and one —the bile, or gall, is a pungent, nauseous bitter. Now, suppose the bile, instead of the saliva, had been secreted by the salivary glands, so that the mouth would be constantly filled with gall, instead of the tasteless saliva.

Mary.—O, Father! that would be *horrid!*

Father.—Well, then, is it not goodness in God, so to arrange the secretory organs as to avoid what "would be *horrid?*" The acrid secretions take place in those parts of the body where they produce no more unpleasant sensation than the saliva does in the mouth. Each secretion of the body is perfectly adapted and agreeable to the part where it takes place; and so far from producing any unpleasant feeling, it is supposed that all the pleasurable bodily sensations we enjoy, are produced by the healthy and faithful action of all the secretory organs. But it is doubtless a truth, that of all the functions of the animal economy, none are more obscure and mysterious than that of secretion; I mean the process by which the fluids or secretions are produced.

The importance of secretion.

Whether the function is mechanical or chemical, has been a subject of controversy among the learned.

Mary.—Father, what do you mean by the function being mechanical or chemical?

Father.—For the function to be mechanical the secretory organs must act like a strainer, separating, or straining out the secretion from the blood as it passes through them.

To be chemical, a change must be produced by the secretory organs, in the elements separated from the blood while passing through these organs.

But experiments have been made, which prove that the function of secretion depends on *nervous* influence.

Mary.—What is the proof, Father?

Father.—The proof is this:—If the nerves which go to the secretory organs be tied with a ligature, so as to cut off the nervous connection with the brain, the function of the organ is at once suspended, and no fluid is secreted.

These secretions are going on during every period of life, and perform the all important office, as I have before said, of removing the various impurities of the blood.

So important is the function of the secretory organs, that when suspended, for even a short time, disease is the sure result. For example:—If from the effect of a chill, the perspiratory glands cease to secrete the

Excretory ducts.

insensible perspiration, fever, or inflammation of some of the internal organs, will follow. If the liver becomes diseased, so as to derange the secretion of bile, either in quantity or quality, digestion is soon impaired. A suppression of any other secretion, deranges the various internal organs, and if long continued, causes death.

The free use of ardent spirits deranges the secretions, and thereby shortens life.

Connected with the secretory glands, are what are called the excretory ducts, which are small tubes, the use of which is to carry off the fluid secreted by the gland,—which act is called *excretion*.

When conversing upon the subject of digestion, I said so much about the secretion and excretion of the *digestive* fluids—the saliva, gastric juice, bile, etc., that I think I need say no more in regard to them now.

There are other extremely interesting facts in regard to secretion and excretion, about which I should like to speak. But as they would be more appropriate for lectures to medical students, than for social converse, I shall defer the consideration of them for the present. I would remark, however, that the facts to which I refer, evince more clearly and forcibly, the kind care and benevolent design of the Creator, than any which we have considered upon the subject. And at some future time, I may think proper to speak freely in reference to them.

Effects should secretion cease.

Before we close our present social interview, I wish to ask you, my children, to reflect, for a moment, upon what would be our condition should all these secretory organs cease their action, for a single day, and thus allow all these various secretions to remain in the blood.

Mary.—What *would* be the effect?

Father.—The blood would become a mass of impurity, causing death in a few hours.

Thus I have given you a very brief description of the process of secretion and excretion, by which you see that the action of the secretory organs, like that of the heart, and many other upon whose action life depends, is going on during every moment of life, without our care;—and, I may add, with no small portion of the community, without even the knowledge that a secretory organ exists in the body, or that such a function as secretion is ever performed.

We must now close this conversation.

Children.—Before we close, Father, we wish to express to you our sincere thanks for the interesting information you have given us.

QUESTIONS ON CONVERSATION XIII.

PAGE 187.—What is here described? Name the most important of these organs or glands. What is their function? What is secreted by the liver?

PAGE 188.—What by the kidneys? What by the salivary glands? What by the lachrymal? What by the pancreatic? What by the perpiratory? What by the synovial? Some of these fluids are designed to do what? What only purpose do others subserve? And still others do what?

PAGE 189.—What is said of the liver? What of the bile? What is said of the kidneys and perspiratory glands? What is done by the kidneys?

PAGE 190.—For what were they designed solely? What depends upon their faithful action? Describe the perspiratory glands. How are they situated? What is the number to the square inch? What is the whole number? Describe the two forms of perspiration. What is connected with the perspiratory glands? What is the form and length of this tube?

PAGE 192.—How is the existence of insensible perspiration proved? Name another proof. What amount of perspiration passes from the body in twenty-four hours? WHat other object is accomplished by perspiration?

PAGE 193.—What is said of experiments made upon the blood, and for what purpose? What did these experiments fail to prove? What else is said on this page, or what inquiries are made?

PAGE 194.—What explanation is given? What is said of the structure of each gland? Each gland being adapted to what? What is hidden from man? What is said of the blood from which these fluids are taken?

PAGE 195.—What marked proof of goodness is there in this? What is said of the acrid secretions? What of each secretion? What is supposed? What truth is stated?

PAGE 196.—What has been a subject of controversy? What is meant by mechanical and chemical? What have experiments proved? What is the proof? What is said of the importance of the function of secretion? Give examples.

PAGE 197.—What deranges the secretion? What is connected with the secretory glands? What is their use? What is the act or function called? What is said of other facts in regard to secretion and exertion?

PAGE 198.—What would be our condition, should the action of these organs cease for a single day? What is farther said of the action of the secretory organs?

CONVERSATION XIV.

ABSORPTION.

CHANGES TAKING PLACE IN OUR BODIES DURING EVERY MOMENT OF LIFE—THE LYMPHATICS, OR ABSORBENTS—DESCRIPTION OF THE LYMPHATICS—ABSORPTION COUNTERBALANCES DEPOSIT, OR NUTRITION—THE GROWTH OF THE BODY, HOW PRODUCED—THE LYMPHATICS NUMEROUS IN THE SKIN, AND MUCOUS MEMBRANE OF THE LUNGS—THIRST ALLAYED BY ABSORPTION—NUTRIMENT ABSORBED THROUGH THE SKIN.

Father.—My dear children, I have spoken in previous conversations of the bones, muscles, nerves, the process of digestion, the formation of the blood, its circulation, the manner in which our bodies are formed, and the process of secretion, by which the blood is kept pure. I now wish to propose a question which either of you are at liberty to answer. It is this:—After the bones, muscles, nerves, brain, and all parts of the body are fully formed, and have arrived at maturity, why do our bodies need nourishment,— or, in other words, why do we need to eat food?

Fred.—That seems a very strange question! I should think any one could answer that. You must think us very ignorant.

Father.—I have not accused you of ignorance. And as you say that any one can answer the question, I hope your reply will prove that the term "ignorant" does not belong to you. So let us have your answer, if you please.

Fred.—We need to eat food because we cannot *live* without it.

Father.—*My son!* If you consider *that* an answer to the question, I do not wonder you thought strange I should ask it. Everybody knows, of course, that we cannot live without eating. But the question is, *why* do we need to eat food, or *why* can we not live without it?

Fred.—Well, Father, if what I have said is not correct, or is not a proper answer, I will confess at once, that I cannot tell, and will leave it to Frank or Mary to answer.

Father.—Well, Frank, can you tell me why we need to eat food after the bones, muscles, and all parts of the body are fully matured, and in full health and strength?

Frank.—We should lose flesh, and become weak and sick, if we left off eating.

Father.—That is true, my son, but the question is, *why* should we lose flesh, and become weak; and in what way, or by what process, can the flesh go off?

Mary.—I think I can answer the question.

Father.—Well, my daughter, I am ready to hear.

Use of nutriment.

Mary.—I recollect once, when we were talking about the body's being nourished by food, you said that all the different parts absorbed, or attracted the nutriment from the blood. Now, if we eat no food there will be no nutriment to absorb. Is not that the reason why we cannot live without eating?

Father.—But, my child, why do the various organs and tissues of our bodies *need* to take up nutriment from the blood, when every part is in a perfect and sound state? Is there need of adding any thing to the bones, when they are already perfect? Or, do we need any addition to the muscles and nerves, or any other parts of the body, when all are perfectly matured, and in full vigor? To add to them in this condition, would be to cause a constant growth of the whole body.

Mary.—I must acknowledge that I do not know, and must ask your explanation.

Father.—I did not propose this question, my children, because I expected you to answer or explain correctly, but rather to impress upon your minds the fact, that there are numerous operations going on in our bodies, upon which life depends, of which a vast majority of the community know and think as little, as they do of what is going on in the moon. It may appear strange to you, that I should require you to speak on a subject of which I know you have not sufficient knowledge to give a correct and intelligble answer. And my only reason for doing so is, that I wish you to *learn to think*,—that is, to take a subject and *study it*, *reflect*, *reason upon it*, and not dismiss it until you are sure that you have arrived at the truth, in regard to it.

Fred.—I wish, Father, you would give us an *example* of how we can reason upon a subject of which we have little or no knowledge.

Father.—As you ask for an example I will give it,—and will take the subject under consideration, and show you how, by close and consecutive reasoning, you could have spoken more intelligibly, and not have been obliged to say, as you have sometimes done,—"I do not know, I never thought on the subject before."

This never thinking upon a subject, is a very mortifying and criminal admission. Thought and intelligence are what distinguish man from the brute. And do you suppose that God has bestowed upon man the powers of intellect, to lie dormant? Do you recollect that God, by the Prophet, said to his chosen people Israel, when complaining of their ingratitude,—"Israel doth not *know*, my people doth not *consider*."

The want of consideration, and the consequent want of knowledge, are assigned, by God himself, as the cause of the ingratitude and wicked conduct of the Israelites. Had they been thoughtful and considerate, they would not have been guilty of such wickedness.

How food nourishes the body.

But we will attend to the example you ask. You know this fact,—for I have explained it,—that the growth of the body, is the result of the deposit of nutriment taken from food and air,—that the body increases in size, by an accumulation of particles, until it obtains maturity, when it ceases to grow. Having learned this fact, your next inquiry should have been, what becomes of the nutriment of food and air (for we still eat and breathe), as it is no longer employed in the growth of the body. This would have led you to the correct conclusion, that, either no particles of nutriment are now deposited in any part of the body, or, if deposited, other particles must be displaced to make room for them; and thus you would have got upon the track of a correct solution of the problem.

Fred.—I feel the truth of what you say, and I will try to make good use of it. But, Father, I wish you would explain this subject to us, for I see, by what you said in trying to teach me to think, that it is one of great interest.

Father.—I will now proceed to explain the subject of our present conversation, which is

ABSORPTION.

Frank.—And what do you mean by absorption!
Father.—I will tell you, after stating an important fact, which is this:—

During every moment of life, there is a change

going on in the particles of which our bodies are composed. Every particle, whether of bone, muscle, nerve or membrane, is a living, vital particle when deposited, and remains so for an indefinite period, and until it subserves a definite purpose, when it loses its vitality, and becomes dead, worn out matter; and if allowed to remain in the body, would prove a source of irritation, and cause disease and death. Now, these lifeless particles, which are in every part and fibre of our bodies,—in the bones as well as in the flesh, must, in some way, be removed, or we could live but a short time. Now, how do you think this can be done?

Frank.—I do not know, Father. Do tell us.

Father.—God has made, in our bodies, a vast number of minute tubes, called *Lym-phat'-ics*, whose function is to absorb or take up this dead, useless matter, and convey it out of the body.

The lymphatic vessels originate in every part of the body, but are most numerous in the skin and mucous membranes, especially the mucous membrane of the lungs. They are so extremely minute, at their origin, as to be invisible, without the aid of the microscope.

Like the veins, and for the same reason, the lymphatics are supplied with valves, and, running together, form tubes larger and larger, as they advance towards the heart.

Frank.—That is wonderful! But what do the

Lymphatic vessels.

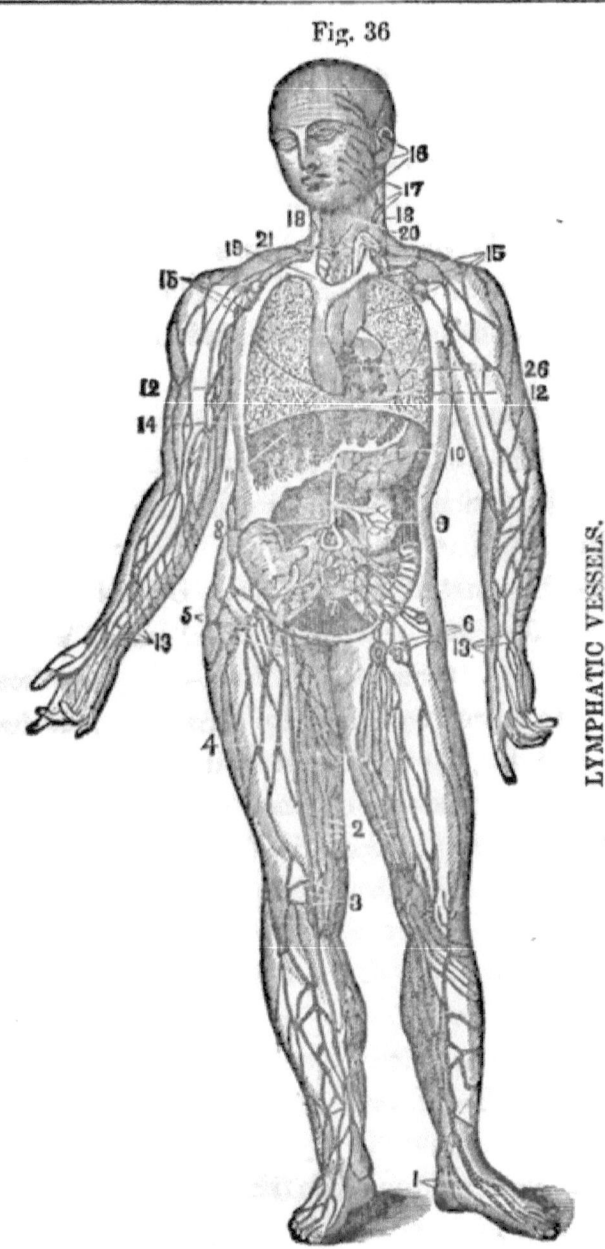

Fig. 36.

LYMPHATIC VESSELS.

Fig. 36. 1, 2, 3, 4, 5, 6, show these vessels of the lower limbs; 7, the inguinal glands; 8, the commencement of the thoracic duct, into which the contents of the lymphatics are poured; 9, the lymphatics of the kidneys; 10, those of the stomach; 11, those of the liver; 12, 12, those of the lungs; 13, 14, 15, those of the arm; 16, 17, 18, those of the face and neck; 19, 20, the large veins; 21, the thoracic duct; 26, the lymphatics of the heart.

Dead matter taken up by the absorbents.

lymphatics do with this worn out, lifeless matter? I hope they do not carry it into the blood.

Father.—They *do* carry it into the blood. It is poured into the large veins near the heart.

Frank.—Well, really, Father, I do not know what to think of that! In our last conversation, you told us about vessels, or organs, made for the very purpose of cleansing the blood, by drawing off the impurities, and I thought it was very fine, and felt very happy to learn that God had provided such wonderful means to purify the blood, as I have heard it said that many diseases are caused by impurities of the blood. And now you tell us, that there are a very great number of vessels in every part of our bodies, taking up dead, worn out matter, and *pouring it into the blood.* I would like to know what use there is in having one set of vessels at work all the time to cleanse out the impurities of the blood, and another set at work pouring impurities into it. This is very strange!

Father.—Do not be alarmed, my son; remember that God's works and ways are perfect, however "strange" they may appear to us; and we have only to study them, and find the *truth*, to make all plain. You will, I doubt not, be ready to admit, that you have been rather hasty in allowing your equanimity to be disturbed, when I inform you that this worn out matter, poured into the veins by the lymphatics, forms no small part of the impurities taken from the blood by the secretory organs.

Absorption equals nutrition.

Frank.—Is that the fact?

Father.—It is.

Frank.—I *do* wish I could learn not to be so hasty in condemning what I cannot understand.

Father.—That is an important lesson for all to learn, especially when investigating or studying the works of God.

An interesting fact in regard to the lymphatic vessels is, that their action of absorption, counterbalances the action of the nutritive vessels; that is, the lymphatics absorb and remove the *dead* matter, as fast as the nutritive vessels deposit the *living* matter, and thus a balance is kept up between the amount deposited and the quantity absorbed—I mean in a state of health. And to this we are indebted for the *uniformity* in the amount of flesh a healthy person possesses, month after month, and year after year. Should the nutritive vessels, as is sometimes the case, (for there are exceptions to this rule), deposit more than the absorbents remove, a person would become plethoric, that is, would become large, by a deposit of fatty matter. If, on the contrary, the lymphatics should be more active than the nutritive vessels, a person would become emaciated.

Mary.—How is that? Do the lymphatics absorb, or take up any thing but worn out matter?

Father.—To absorb the worn out matter, seems to be the primary object of the lymphatics, but they have other, and important functions to perform, as I shall show you before we close our present interview.

Fred.—Father, you say, that in a healthy state, the absorption equals the deposit, and in this way there is a balance kept up between the amount of matter deposited, and the quantity absorbed. Is this the case all through life?

Father.—No, my son, it is so only after a person arrives at maturity—that is, after the body is fully grown; and after that, it is so *only* while the individual is in a state of health.

During the periods of infancy, childhood, and youth, the nutritive vessels are more active than the absorbents; consequently the deposit of matter in every part of the body, is greater than the absorption, which causes a gradual increase, or growth of the body. But after a certain number of years,—varying in different persons,—the nutritive vessels become less and less active, until the deposit and absorption become equal, when the growth of the body ceases;—after which, if the person is healthy, and temperate in eating and drinking, the bulk and weight of the body remain about the same till advanced life, when the balance between the two being lost, and nutrition becoming less than absorption, the body gradually diminishes in size and weight, and finally presents the shrivelled appearance peculiar to the aged.

Fred.—I never knew before in what way our bodies grow; nor why old people are so thin and poor.

Matter composing the body changed every seven years.

Father.—I would here remark, that the absorption which is going on every moment of life, produces an entire change in the particles of which our bodies are composed, every seven years; that is, every atom of matter constituting the body seven years ago, has been removed by absorption, and their place supplied by a new deposit, so that our bodies do not now contain a particle of the matter, either of bone, muscle, nerve, or membrane, of which they were composed seven years since; and some physiologists believe the change takes place in a much shorter period.

Fred.—Do we then have new bodies every seven years?

Father.—We do, so far as the matter is concerned, of which they are composed. And yet, we have the same identical body all through life; our friends never mistake us for another person. Nor was the son of the Emerald Isle correct, who claimed that he was a native American citizen, because he had lost the old body which he brought from Ireland, seven years ago, and had obtained a new one in America.

You know, my children, that persons often become greatly emaciated, that is, lose their flesh, during sickness. Did you ever think how it is possible for this to take place?

Mary.—I only thought it was because they were sick. I can give no other reason. I cannot tell *how it is done.*

Action and value of the lymphatics.

Father.—When persons are so sick as to lose flesh, as it is called, the digestive organs are not in a condition to digest food. But a person cannot live long without nutriment, sick or well. What then is to be done, when the digestive organs refuse their service in preparing nutriment? I will tell you what *is* done. The lymphatics absorb the fatty and other matter which has been deposited in various parts of the body, and carry it to the blood, for food or nourishment to the system. Thus persons, in such case, *live upon their own flesh.*

Mary.—Do you mean, Father, that healthy portions of the body are taken up by the lymphatics and conveyed into the blood, for the purpose of nourishing the system?

Father.—Yes, my child, that is the fact. There are deposits, in various parts of the body, of adipose matter, which are not needed to sustain life or health, and which seem to be stored away, to be used in case of emergency. In "fleshy" persons, this deposit is great, and in some cases very burdensome. Now, this matter, which often lies stored away for years, can all be taken up by the absorbents, conveyed to the blood as fast as needed, and used by the system as nutriment.

In case of shipwrecks upon the ocean, the life of mariners is sometimes sustained by this means, for weeks. Is not this a wonderful provision of our kind heavenly Father?

Thirst allayed by the absorbents.

Mary.—It certainly is, but how ignorant I have always been about these things. I feel grateful to you, Father, for consenting to give us these instructive lessons.

Father.—There are, also, secretions and deposits in the body which have no external outlet. These are the deposits of marrow in the bones, the secretion of the synovia, or joint-water, and several others. Were it not for the absorption of the lymphatics, these deposits and secretions, would soon be in excess, and dropsy of the parts would be the result. But now, in a healthy state, the absorption equals the secretion and deposit, therefore these fluids are neither in excess nor deficient,—neither too much nor too little.

I have told you that the lymphatics are very numerous in the skin. So numerous and *active* are these absorbents that thirst may be allayed by going into a water-bath. And sick persons, unable to take food, by being immersed in a bath of broth, may absorb, through the skin, a sufficient amount of nutriment to sustain life for some time.

Substances applied to the surface, such as Spanish flies, mercury, morphine, etc., are often taken into the system, through the skin by the action of the lymphatics, and produce the same effects as if taken by the mouth.

In case of a *bruise* upon the surface of the body, by

Action of the lymphatics.

which the minute blood-vessels are injured, the blood coagulates and causes the part to present a dark color—sometimes quite black—especially is this the case if the injury be on the face near the eyes.

Now, if this coagulated blood were allowed to remain, it would become putrid and cause ulcers. But the lymphatics, faithful to the discharge of their duty, absorb it, and the parts gradually assume their natural color.

I have stated that the mucous membrane of the lungs is abundantly supplied with lymphatics. The consequence is, that substances inhaled are quickly taken up and conveyed into the circulation. It is in this way that sulphuric ether, taken into the lungs by inhalation, is rapidly conveyed to the brain, and thus produces its effect upon the nervous system. Substances that are injurious, as well as beneficial, may be absorbed by the lymphatics.

Thus, contagious diseases may be contracted, by inhaling into the lungs, and absorbing through the skin, the effluvia containing the contagion.

A knowledge of this fact may enable us to avoid contagious diseases, by avoiding the impure air containing the contagion, and by general cleanliness of person and clothing.

In case of disease of the lungs, great benefit may be derived,—and it is affirmed by some, that cures have been effected, by inhaling medicated vapor.

Reflections.

There are other interesting facts in regard to the absorbent vessels, but we have not time to pursue this subject further.

From the facts stated we see, that in a *physical*, as well as moral sense, " *God careth for us.*"

What wisdom in design,—what perfection in structure, do our bodies present! How numerous the proofs, that God intended to mingle pleasure with physical existence!

The evidence of benevolent design that meets us on every side, as we pursue our investigations, is but an echo of the inspired declaration,—" *God is love.*"

QUESTIONS ON CONVERSATION XIV.

PAGE 199.—Why do our bodies need nourishment after they have arrived at full maturity? What is said on pages 200, 201, 202?

PAGE 203.—Of what is the growth of the body the result? How does the body increase in size? What more is said? What is the subject of this conversation?

PAGE 204.—What important fact is stated? What must be removed? How is this done? Where do the lymphatics originate? Where are they most numerous? What is said of their size? With what are they supplied? What do they form?

PAGE 206.—What is done with this worn-out matter? How is it removed from the blood?

PAGE 207.—What interesting fact in regard to the lymphatics is here stated? What is thus kept up? For what are we indebted to this?— What is farther said of the action of the nutritive vessels and lymphatics? Do the lymphatics take up anything but worn-out matter?

PAGE 208.—Does the absorption equal the deposit all through life? What is said of the nutritive vessels during the period of infancy, childhood, and youth? What is the result? What takes place after a certain number of years? Until what? What remains the same? What then takes place?

PAGE 209.—What change is produced by absorption? Do we have the same body all through life?

PAGE 210.—In what way can persons become emaciated or lose their flesh? What is said of deposits in various parts of the body? What in case of shipwreck?

PAGE 211.—What is said of secretions and deposits which have no outlet? What is said of the lymphatics in the skin? What are often taken into the system?

PAGE 212.—What is done by the lymphatics in case of a bruise? The lymphatics of the lungs do what? What is said of contagious diseases? What may be done in case of disease of the lungs? What do we see in these facts?

CONVERSATION XV.

THE SKIN.

THE GREAT NUMBER AND VARIETY OF THE VESSELS OF THE SKIN—DESCRIPTION OF THE SKIN—THE CUTICLE—THE DERMA—THE COLORING MATTER—THE NERVES AND BLOOD-VESSELS OF THE SKIN—INTERESTING FACTS IN REGARD TO THE CUTICLE AND DERMA—THE OIL GLANDS, THEIR FUNCTION—THE SKIN A BREATHING APPARATUS—THE PORES OF THE SKIN, THEIR FUNCTION—THE NUMBER OF PORES—THEIR AGGREGATE LENGTH—REFLECTIONS.

Father.—My children, I wish to converse with you, at the present time, in regard to a part of our bodies which contains a greater number and variety of vessels, and fulfils a greater number of important purposes in the animal economy, than any other on which we have yet conversed;—I refer to the external covering of the body—the SKIN.

Fred.—Are the vessels of the skin more numerous than the blood-vessels, which, I recollect you said were millions?

Father.—I shall show you, before we close our present interview, that no organ, or membrane of the body, is composed of so many vessels, as the membrane that envelops the body. Could you view one

Vessels of the skin.

square inch of the human skin, while in a state of health and activity, with a powerful microscope like the philosopher Humboldt's, which, it is said, magnified objects three hundred thousand times, you would be frightened at the sight, and would start back with instinctive horror.

Frank.—What should we see, Father?

Father.—The first that would meet your view, probably, would be great numbers of *vast rivers of blood*, lying close together, yet never mingling, because enclosed in tubes. These tubes are the arteries, veins, and capillaries. You would also see a great number of canals, of great magnitude, whose mouths open upon the surface, and pour out rivers of fluids. These canals are the pores of the skin. You might next observe a dense fog, rising like steam from the boiler of a steam-engine. This fog is the insensible perspiration, which rises from, and envelops the whole body. You would also see a perfect network of nerves, too minute to be seen with the naked eye, but now appearing like branches of large trees.

Mary.—O, Father! that would be a frightful sight. But I like to hear you talk about them.

Father.—There are many interesting facts in regard to the skin, with which I wish to make you acquainted; but that you may the better understand, I must be a little systematic. The skin is a smooth, delicate membrane, forming a most beautiful covering for the whole body, and, as one justly remarks, "is the last stroke of the great Artist, which gives the finishing touch, and makes the form divine."

To the naked eye, the skin appears to be but one membrane, but a careful examination will show that it is composed of two layers of membrane, the *Cu'-ti-cle*, (scarf skin), and the *Cu'-tis Ve'-ra*, or *Der'-ma*, (true skin).

Fig. 37.—THE SKIN, MAGNIFIED.

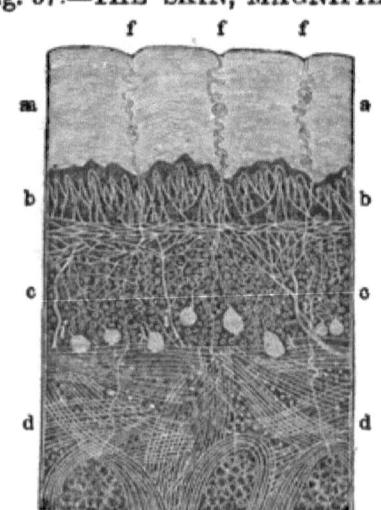

Fig. 37. a, a, the cuticle; b, b, the colored layer: c, c, d, d, the true skin; e, e, e, fat cells; f, f, f, sweat tubes.

Between these two layers of membrane, is a thin layer of paste, called *Re'-te Mu-co'-sum*, (mucous coat of skin), which gives *color* to the skin. In our race, this layer of paste is white, in the Spanish, it is yellow, in the Indian, it is copper colored, and in the African it is a jet black.

Of what the skin is composed.

The Derma, or true skin, is composed of Arteries, Veins, Capillaries, Nerves, Lymphatics, Oil-Glands and their Tubes, and Perspiratory Glands and their Tubes.

The arteries, which are very numerous, divide and subdivide into myriads of minute capillaries, which form a beautiful and perfect net-work on the surface of the true skin, and terminate in minute veins, which are as numerous as the arteries, as I have previously explained. So perfect is this net-work of blood-vessels, that the point of the finest needle, introduced into the skin, is sure to wound one or more of them.

In regard to the *nerves* of the skin, I can only repeat what I have previously stated,—that they cover the entire body, and are extremely sensitive to the touch.

Fred.—Why are the nerves in the skin more sensitive than in other parts of the body?

Father.—For the reason, as I told you in a previous conversation, that the surface of the body is more liable to injury than the internal parts, and, therefore, needs to be more carefully guarded. Had there been left upon the body, an extent of surface without nerves, equal in size to a dime, a sharp pointed instrument might pierce that spot, so as to endanger life, and yet cause no sensation, and therefore give no warning of danger. But there is no such imperfection in the works of God. He saw that every particle of the surface of our bodies requires to be guarded from danger, and therefore he has placed a sentinel, (nerve), at every point to give instant warning of an approaching foe, as I have already told you. But in the design and formation of the human body, there was a most important matter to be arranged, in regard to these sensitive nerves of the skin. It was necessary that the skin should be sensitive, in order to give warning of danger, but as the surface of the body, in the common duties of life, comes in contact with surrounding objects, it was necessary also, that this nervous sensibility should, in some way, and to some extent, be modified, or rendered moderately acute,—otherwise, every thing that touched the body would cause pain, and thus we should be kept in a state of constant suffering.

In addition to our *safety*, which was to be provided for, was another consideration of, perhaps, no less importance. I refer to the fact, that all the knowledge of external nature, obtained by the sense of feeling, is conveyed to the mind through the medium of the sensitive nerves of the skin. Had these been too sensitive, every thing we touch would cause pain, instead of conveying intelligence to the mind. Had the sensation of touch been too little, the nerves would fail to be a medium of intelligence between the mind and external nature, and consequently the benefit to be derived from the sense of feeling would be lost.

Now how can both these objects be attained? That *is*, how can the nerves be rendered sufficiently sensitive to convey correct information to the mind, relative to whatever we touch, and at the same time, not be so sensitive as to cause pain.

Fred.—I cannot tell, Father. Do please explain it to us.

Father.—The course adopted by the Creator, to prevent a too great sensitiveness on the one hand, and too little on the other, is to cover the derma, or true skin, with a thin, soft membrane, destitute of nerves, and consequently without sensation.

This important membrane is what is called the cuticle, or scarf-skin. It seems perfectly adapted to the function it performs; for, while it moderates the sensibility of the nerves, so that contact with external objects causes no pain, it, at the same time, allows the degree of sensation requisite to enable the nerves to take cognizance of whatever comes in contact with the surface of the body.

Mary.—How wonderful!

Father.—There is one very interesting fact in regard to the cuticle, which seems to me to be a special design of the Creator, for our comfort and safety. You know that the more cloth, or leather, is worn, the thinner it becomes, until it is worn out. But the *reverse* is true of the cuticle. The harder the use, and the greater the wear of the cuticle, the *thicker* it becomes, which is proved by the hard and callous appearance of the hands of farmers, black-smiths, masons, and mechanics.

Fred.—That seems very strange! How do you explain it?

Father.—This thickening of the cuticle of the hands, which enables workmen to handle their tools and materials without pain or inconvenience, was a matter of necessity; for if the wear of the *cuticle* caused it to become thin, as it does every thing else, it would soon wear out, on the hands of the laboring man, leaving the nerves bare, which, of course, would utterly disqualify him for labor. God saw that this condition of the cuticle was necessary, and therefore endowed the skin with the requisite power to produce it, and that is all the explanation I can give of this wonderful phenomenon. In other words, it is so because God " careth for us."

In a previous conversation, I have spoken of the lymphatics of the skin, also, of the perspiratory glands, and their excretory tubes, and I need not repeat what I then said. There remains one other set of vessels of the skin to be described, which are called the *Se-ba'-ceous Fol'-li-cles*, or oil-glands, which are small bodies in the true skin, and are connected with small tubes which pass out through the cuticle.

The oil-glands are not spread equally over the whole surface, indeed, they are entirely wanting, in some

Oil-glands and their secretions.

Fig. 38.

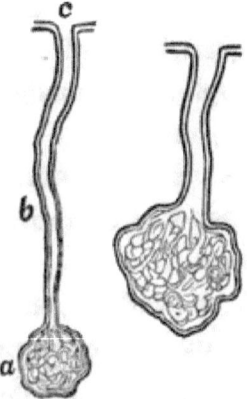

OIL-GLANDS AND TUBES, MAGNIFIED.
Fig. 38. a, the gland; b, the tube; c, its mouth.

parts of the body, while they are very abundant in others, where their office is most needed, as on the face, the head, the nose, the ears, etc.

Fred.—I wish to know what is meant by oil-glands!

Father.—These are glands which secrete an oily fluid, which is a compound of oil and albumen, as it is called.

Fred.—But what is the object of such a secretion?

Father.—The object seems to be two-fold.—1st. This oily fluid serves the purpose of keeping the cuticle soft and smooth, and thus prevent its becoming parched, and causing chaps and sores. 2d. To remove waste matter from the system. And it is worthy of remark, that the oil-glands are placed in those parts of the skin which are most exposed to changes of temperature and moisture of the air, as the face and head; while in those parts protected from the air by clothing, and in which, of course, the cuticle is kept moist and supple by insensible perspiration, few or none of the oil-glands exist. They are also found in the parts liable to chafing, where they pour their oily secretions upon the surface, and thus prevent friction and irritation.

Frank.—I did not think the skin had so many, and such important functions to perform.

Father.—I have not named them all. Added to the others, the skin is an organ of respiration—or a breathing apparatus.

Frank.—What is that? I thought we breathed with the lungs.

Father.—So we do,—and we breathe with the skin also, as breathing consists in absorbing oxygen from the air, and throwing off carbonic acid by the lungs; the same thing is done by the skin; the absorbents of the skin of which I have previously spoken, taking up oxygen from the surrounding air, while the carbonic acid is exhaled, or thrown off through the pores of the skin.

Fred.—I should like to know how that is proved.

Father.—It is a very easy matter to demonstrate this position, and it has been done repeatedly. Inclose the body, or a part of it, with a quantity of air, in an India rubber, air-tight case, and, in a few hours, it

Proofs that we breathe through the skin.

will be found, that the air in the case has lost its oxygen, and is filled with carbonic acid,—the oxygen, of course, having been absorbed by the body, while at the same time it had thrown off the carbon.

Mary.—How can it be known, Father, that the air in the case has lost its oxygen?

Father.—It can be ascertained in various ways, but the following is, perhaps, the most ready and satisfactory.—Place a burning lamp in the case containing the air, and it will go out instantly.

Mary.—How does that prove it?

Father.—Do you not recollect I have already told you, that combustion, or the burning of any thing is produced by the chemical union of oxygen with carbon; that the burning of coal or wood is the result of a union of oxygen with the carbon contained in the fuel?

Mary.—I recollect that, but I do not understand how it proves anything in this case.

Fred.—I see how it is,—please let me explain.

Father.—Certainly, my son, if Mary cannot.

Fred.—If the oxygen had been absorbed from the air which surrounded the body in the rubber case, there was none to unite with the carbon of the oil, and of *course*, a burning lamp, placed where there was no oxygen, would go out. Is that right, Father?

Father.—That is correct, my son, and is very well explained. I trust now you are all satisfied that the skin actually breathes, as well as the lungs, though in a different way. And what a thought, that our bodies are every moment absorbing *life* from the air around us and within us!

Frank.—I think we understand it, Father, and it is wonderful!

Fred.—You have spoken several times, about the pores of the skin, but you have not told us what they are.

Father.—The pores are the outlets or mouths of the tubes connected with the perspiratory glands and the oil-glands. These tubes, which are spiral in form, are about one-fourth of an inch in length, and act as drains to carry off the perspiration as fast as secreted, as I have before stated. (See Fig. 35, page 191). From what I have said in regard to the number of perspiratory and oil-glands, and their tubes,—the mouths of which, as before stated, are the pores,—you will see that the number of pores must be very great.

Fred.—Can you tell us how many pores there are?

Father.—Dr. Wilson, the author of a work on anatomy says, that, with a view to ascertain the number of pores, he counted, (with the aid of the microscope, of course,) the pores on the palm of the hand, and found three thousand five hundred and twenty-eight to the square inch. But this, he thinks, is above the average number of the body, which he estimates at two thousand eight hundred to the square inch. Now in a man of ordinary height and bulk, the number of square inches of surface is about two thousand five hundred—the number of pores, therefore, is *seven millions*,—and the number of inches in length, of the perspiratory tubes, is one-fourth of that number—one million seven hundred and fifty thousand, which is one hundred and forty-five thousand, eight hundred and thirty-three feet, or forty-eight thousand six hundred and eleven yards, or nearly *twenty-eight miles.*

What a thought! *seven millions* of tubes for *drainage*, on the surface of the body! Now, as God makes nothing in vain, these seven millions of pores must be necessary for the enjoyment of health and the continuance of life. What then would be the effect should this drainage be obstructed?

Mary.—What *would* be the effect, Father?

Father.—Disease and death would be the sure and speedy result.

Mary.—I do not wonder that the Psalmist says,—"I am *fearfully* and wonderfully made." I almost fear to move, lest in some way, I should interrupt the operations of the millions of vessels on which life depends.

Father.—Then, my child, you can appreciate the feelings of the immortal Watts, when he exclaims,—

> " Our life contains a thousand springs,
>   And dies if one be gone,
> Strange that a harp of thousand strings,
>   Should keep in tune so long!"

What the skin hides from view.

The skin fulfils the important purpose of covering up, or concealing from view, the operations that are constantly going on within the body. Could we see through the skin, and the integuments under it, the complicated mechanism of our bodies in full operation,—could we see the heart in constant motion, pulsating eighty times a minute, and at each pulsation throwing a jet of two ounces of blood from each ventricle, into the arteries,—the blood rushing through the arteries at the rate of sixty feet per minute,—the secretory glands drawing off their various secretions from the blood,—the muscles and tendons pulling the limbs in every direction,—the lungs inhaling and blowing out air;—and the numerous minor, but not less important operations that are going on in all parts of the system, we should be *paralyzed with terror, and should stand in awe of ourselves!*

But all these are kindly hidden from our sight, and, to most persons, are no more subjects of observation, or even of thought, than the operations that are going on within the volcanoes of the moon.

We must now close this pleasant interview, with one reflection, from which we shall learn a moral lesson. From what has been said we see that the millions of minute vessels in our bodies, are formed with as much skill, and preserved with as much faithfulness, as the larger—and what may appear to us, the more important organs. Perfection, faithfulness,

Reflections.

and goodness, mark the most minute, as well as the most magnificent, of God's works;—by which we should learn, that to be like him, we also must be faithful in that which is least, as well as in the more important actions of our lives.

QUESTIONS ON CONVERSATION XV.

PAGE 214.—What is said of the vessels of the skin?

PAGE 215.—What is to be seen in a square inch of the human skin greatly magnified? Describe the skin.

PAGE 216.—Of what is the skin composed? What are these membranes called? What is between the scurf-skin and the true skin? What is it called? What is its function? In what race is it white? In what yellow? In what copper colored? In what a jet black?

PAGE 217.—Of what is the derma or true skin composed? What is said of the arteries?—What of the capillaries and veins? What is said of the network of blood-vessels? What of the nerves of the skin?

PAGE 218.—In the design and formation of the human body, what important matter was to be arranged in regard to the sensitive nerves of the skin? What other fact is referred to on this page? Describe it fully.

PAGE 219.—How can both these objects be attained? What course has the Creator adopted? What is this important membrane? It does what? What other interesting fact in regard to the cuticle is named?

PAGE 220.—What was a matter of necessity? With what did God endow the skin? What other set of vessels of the skin is described?

PAGE 221.—What is said of the oil-glands? Describe them, and their function. What is the object of such a secretion? What is worthy of remark?

PAGE 222.—What other important office does the skin perform? In what way do we breath with the skin? How is this proved?

PAGE 223.—How can it be shown that the air has lost its oxygen? What is the most satisfactory proof?

PAGE 224.—What are the pores of the skin? What is the form and length of these tubes? They act as what?

PAGE 225.—What is the average number of pores to the square inch? How many square inches are there on the surface of the body of a man of ordinary bulk? The number of pores, therefore, is what? What is the aggregate length of the perspiratory tube? These seven millions of tubes are for what purpose? What would be the effect upon health should this drainage be obstructed?

PAGE 226.—What other important purpose does the skin fulfil? What operations are concealed from view by the skin? What if we could see them? What is said of the minute vessels? What should we learn by this?

CONVERSATION XVI.

THE TEETH.

THE INFANT BORN WITHOUT TEETH—TWO SETS OF TEETH PROVIDED IN THE JAW BONES BEFORE BIRTH—FIRST SET TEMPORARY, SECOND SET PERMANENT—GOD'S GOODNESS IN PROVIDING FOR A FUTURE WANT—THE NUMBER AND NAMES OF THE DIFFERENT DIVISIONS OR KINDS OF TEETH—FORM AND USE OF THE DIFFERENT KINDS—COMPOSITION OF THE TEETH—THE ENAMEL OF THE TEETH—THE ROOTS OF THE FIRST SET ABSORBED—MOVEMENTS OF THE JAW IN MASTICATING OR CHEWING FOOD—CAUSE OF TOOTHACHE—NERVE AND BLOOD-VESSELS OF THE TEETH—THE TEETH AN ORNAMENT.

Father.—Did it ever occur to you, my children, that there was any thing relating to the *teeth*, that afforded striking proof of the wise design and goodness of our Creator?

Fred.—I never thought of any thing of special interest about the teeth, Father, but judging from the past conversations, I doubt not, if you undertake it, you will tell much, both to amuse and instruct us, which we are quite ready to hear. I do not know how it is with Frank and Mary, but as for myself, I have come to the conclusion, that every part of our bodies, however minute, affords the clearest evidence of the goodness and benevolent design of God, if we can only take the right view of it. Before we commenced conversing upon this subject, I could not have believed it possible, that so many, and such varied proofs of God's goodness could be produced by an examination of the structure of the human body.

Father.—You are certainly right, my son, in your conclusion that every portion of our physical structure bears unequivocal proof of God's goodness in its formation, if, as you say, we only take a right view of it.

I wish to make you acquainted, at the present time, with what to me appears very interesting, with regard to the teeth.

The first to which I wish to call your attention, is, that we are born without teeth.

Frank.—I know we are, but is there any particular and wise design in that?

Father.—I think so, Frank.

Frank.—It seems to me, Father, that you can discover design and goodness where no one else would think of it. I wish I had the same faculty.

Father.—The infant has no teeth, for the reason that it needs none. It is destined to obtain its food from its mother's breast, and so far from teeth being a benefit, they would be a great inconvenience, for how could an infant nurse with a mouth full of teeth? But the time will come when it will need more solid food, and such as will require the aid of teeth to masticate. The Creator knew this, and therefore provided teeth in the bones of the jaw, which, about the sixth or eighth month—just the time when needed, make their appearance, by coming through the gums, as white as ivory, and as sharp as a knife—and by their appearance seem to say,—" Here we are, for by this time you need us."

Now, just think of this, my children. Here is an instance in which God provides for a future want. The infant, at birth, as I have said, needs no teeth, but will need them at a future time. The Creator, therefore, places the germs of teeth, if I may so speak, in the bones of the jaw, which, by growth, make their appearance at the precise time they are needed—first, the sharp, cutting, then, as they are required, the large teeth, for chewing or grinding the food. Now, was there any good and wise design in this?

Frank.—It certainly seems so, Father, but I think I should not have discovered it myself.

Fred.—I have been listening to what you have said, and no doubt it is so; but I have an idea which I should like to have you reconcile with the goodness you have alluded to.

Father.—Well, my son, what is it?

Fred.—These teeth, it seems to me, are not worth much, after all; for they soon get loose and come out. What do you say to that, Father?

Father.—What do I say to that? Why, I say it is additional proof of God's goodness.

First and second sets of teeth.

Fred.—It may be so, but I should like to know what you have to say in proof of it.

Father.—At the early age when the teeth appear, the bones of the jaw are small and tender, and are not capable of holding large teeth, such as men and women have. Still, the little child needs teeth as much as it will when it arrives at manhood. Now, what should be done in such a case? To have large, solid teeth that will last through manhood and old age is impossible. And if it were possible, how would an infant, or a little child appear with a mouthful of large teeth, such as adults have?

Fred.—Would they not grow large as the child grew up?

Father.—No, there is not room for them, consequently they would crowd upon each other. Now were such a case submitted to an ingenious mechanic, what would he say would be the course dictated by wisdom?

Fred.—What *would* he say, Father?

Father.—He would probably tell you, that, as the bones of the jaw are small and tender, and not capable of holding large teeth, there should be a temporary set of small teeth,—such as will answer the present purpose, until the jaw bones grow large and strong, and are capable of holding large teeth; then, the temporary teeth should be taken out, and large solid ones put in, that will last for life. And this is precisely the plan adopted by our Creator.

He has not only provided in the jaw, long before they are needed, the little teeth that come out first, but underneath them, he has also provided another set, which are to become the permanent teeth. This second set of teeth grow up under the first, and at the age of from six or seven, to twelve or fourteen years, they displace these first teeth, as I will soon explain, and grow up in their place. The first teeth, however, are not all removed at once, for that would leave the child without any to chew its food with,— but one or two at a time, disappear, until all are finally gone, and their place supplied by the large, solid teeth, which are designed to remain for life.

The providing of these second teeth, is more wonderful than the first. The first will be needed in a few months, but the second will not be required for several years; yet they are placed in the jaw before the child is born—a treasure, locked up for future development and use, although not necessary for some eight or ten years afterwards. Now what do you say to that, Frederick.

Fred.—I acknowledge my ignorance.

Father.—The only way to improve and become wise, my son, is to be willing, as I see you are, to confess your errors.

Fred.—I am very willing to do that, Father, for the sake of gaining useful knowledge, and am very thankful for the instruction you have given us. And if you have any thing more to say about the teeth, we should be delighted to hear it.

Father.—There are several facts of interest which I wish to name, but really, we are proceeding with very little order or system. Let us be a little more definite and systematic.

The first set of teeth, of which I have spoken, is called the *tem'-po-ra-ry*, or milk teeth. There are twenty of the temporary teeth,—ten in each jaw. The second set are called *per'-ma-nent* teeth. As the jaw bones have become large before the second teeth appear, it takes a larger number of the second than it did of the first to fill the jaw. Accordingly we find that there are thirty-two of the permanent teeth,—sixteen in each jaw.

Fig. 39.

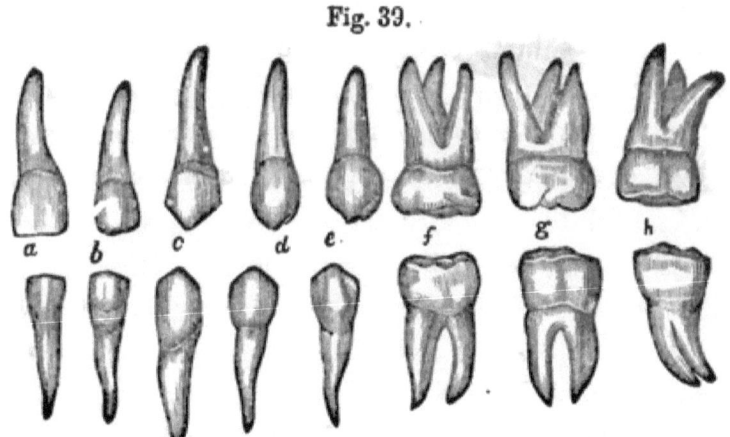

THE TEETH.

Fig. 39. a, b, are the cutting teeth (incisors) of the upper and lower jaw; c, is the eye tooth, (cuspid); d, e, are small grinders, (bicuspids); f, g, h, are grinders, (molars.)

Crown and root of the tooth.

NAMES OF THE TEETH.

The four front teeth, in each jaw, are called *in-ci'-sors*, (cutting teeth); the tooth next to the incisors on each side, is called the *cus-pid* (eye tooth); the next two, *bi-cus'-pids*, (small grinders); the next two, *mo'-lars*, (grinders). As the last tooth on each side does not make its appearance until the person is about twenty years of age, they are called *wisdom* teeth.

Each tooth consists of two parts, called the *crown* and the *root*. The root is firmly set in the socket. The crown is the part that protrudes from the jaw-bone. The incisors, cuspids, and bicuspids, have each but one root. The molars of the lower jaw have two roots, while those of the upper jaw have three.

The teeth of each division that I have named, differ greatly from each other in form and in the office they fulfil, each being perfectly adapted to its peculiar office. The incisors, or cutting teeth, particularly in the upper jaw, are wide, thin and sharp, like a chisel, and are designed for cutting food, such as bread, fruit, etc., and how admirably are they adapted to this object. Take an apple, for example. How easily will one bite a piece from it, the cutting teeth acting like so many chisels. Having separated a piece from an apple, or from a slice of bread, or from any other food, the next object is to grind it, or break down its more solid portions. But this cannot be done by the cutting teeth,—if you should attempt it, you would

Design of the different teeth.

find that you only cut the food into smaller pieces, without grinding or breaking its texture. To accomplish this necessary object, large teeth, with a broad end or surface are needed, and the molars, or grinders, are perfectly adapted to this purpose, and were specially designed for it by the Creator.

The incisors, or front teeth, then, **are for cutting** the food, and the molars, or jaw teeth, as they are commonly called, are for grinding it.

That you may more clearly see the value of this arrangement, and the wisdom and goodness that directed it, let us suppose it to be reversed, and the teeth transposed, so that the cutting teeth would be on the sides of the jaw, and the molars or grinders in front.

Now with these broad teeth in front, could you bite a piece from an apple, or could you grind it with the sharp cutting teeth on the sides? Do you not see that with this arrangement, the teeth would be almost useless and utterly unfit for the office they are to perform?

Mary.—I see, Father, that it is so; and it is certainly wonderful, that God should take such care in making our teeth, so that they should be just what are needed.

COMPOSITION OF THE TEETH.

Father.—The teeth are similar, in composition, to other bones, and yet they are different in some important particulars. They are much harder than bone, and as they are designed for a very different purpose, they require solidity and strength, which qualities they possess in a remarkable degree.

You recollect I told you, when describing the bones, that they were covered with a firm membrane called pereosteum. But the teeth have a covering totally different. It is a hard, smooth, glassy substance, called *en-am'-el*. It is remarkable for its hardness. Few metals are as hard as the enamel of the teeth. The internal part of the tooth is called the *i'-vo-ry*, and forms the body of the tooth.

Frank.—Why do the teeth have a covering so different from the bones.

Father.—For two reasons. 1st. The teeth have a great deal of work to do in grinding the food; and if the outside, or covering, was not extremely hard, they would soon wear out. 2d. The ivory, or body of the tooth, is composed of substances which are readily decomposed by strong acids, while acids have little effect upon the enamel. Now had not the ivory of the tooth been protected by a covering upon which acids have little or no effect, vinegar, and other acids used with food, would soon decompose the body of the tooth, and it would crumble to atoms. This is proved by the fact, that when the enamel, from whatever cause, is broken, the tooth soon decays.

Frank.—I understand it now, and it all shows the kind care of our Creator.

Roots of first teeth absorbed.

Fred.—Father, I wish to ask a question.

Father.—It is usually easier, Frederick, to ask questions than to answer them. But so far as I am able, it will give me pleasure to answer any query you may present.

Fred.—I recollect when my first teeth became loose, and came out, they had no roots; and appeared as if they were kept in place by the gums. Now I would like to ask how they were held in so fast without roots?

Father.—The first teeth have roots as well as the second, and it is the root alone that keeps them in place.

Fred.—How then do they come out without roots?

Father.—As the second or permanent teeth begin to grow and press upon the roots of the first, the root is gradually removed by *absorption*, to make room for the second, the absorption removing the root of the first, as fast as the growth of the second requires it, until it is all absorbed, at which time the second tooth has nearly made its appearance. When the root is all absorbed, there is nothing to hold the tooth but the gums, and consequently it becomes loose.

Fred.—But what can be the object, in the absorption of the root! Can you discover any benevolent design in that?

Father.—It is not to be supposed, my son, that we can always discover the designs of the Creator, by the

Absorption of the roots—a wonderful provision.

limited examinations we are able to make of his works. Man is too ignorant, even the most learned, to assume that, without arrogance. But in this case, it seems to me, that two objects at least, are apparent. Had the first teeth been pressed out as soon as the second began to grow, as they would have been, had not the root been absorbed, the places would have been vacant, or without teeth, for a long time,—until the second had grown out,—whereas, by absorbing the root of the first, the second is allowed so full a growth before the first is removed, that the vacancy is soon filled, and the child is deprived of the teeth but a short time.

But another, and more important reason is the following. Had not the root of the first tooth been absorbed to make room for the second, the first would be pushed out of its socket as fast as the growth of the second required, and would thus protrude beyond the other teeth,—which would prevent their shutting together, and render the mastication of food impossible.

Now, what a condition is this for one to be in,—to have, for several years, some of the teeth projecting beyond the rest, so as to render it impossible to close the mouth. And yet, this would have been the state of every child, had not God in his goodness prevented it, by endowing the absorbents with power to remove the roots of the first teeth as fast as the growth of the second requires it.

Fred.—O, Father, how wonderful all this is!

Father.—It *is* wonderful, my son, and nothing but God's goodness could have prompted such an arrangement.

Fred.—What a sad condition that would be,—to have the mouth kept open in consequence of one or two teeth being longer than the others.

Father.—It would have been dreadful indeed, and let us thank God for his tender care over us, and his goodness manifested towards us.

The teeth, placed in the jaw-bone, present a peculiar appearance. They are set in sockets called *al'-ve-o-lar* processes, which give them great firmness, the tooth appearing like one bone driven into another.

There are two movements of the lower jaw in masticating or chewing food. One is the action by which the mouth is opened and closed, or by which the teeth are brought together when they have been separated by opening the mouth, the other is a lateral or side movement,—the jaw moving from right to left, or from left to right. The first of these movements is for cutting or dividing the food, the latter for grinding it between the large teeth.

The first movement—that of opening and closing the mouth, is produced by two large muscles situated on each side of the face.

The lateral, or grinding movement, is caused by muscles attached to the lower jaw, on the inside. The action of these muscles can be plainly felt, by placing the fingers, while masticating food, upon the face above the angle of the lower jaw. These muscles are of great strength, and are capable of acting with an incredible degree of force. In the disease called locked-jaw, these muscles, from some unnatural irritation, become so firmly contracted that the patient cannot open his mouth, and it would be difficult, in some cases, even to pry it open.

Mary.—I wish to ask, Father, what causes the toothache.

Father.—In the centre of each tooth, is a small nerve, which is entirely destitute of sensation so long as the tooth remains sound; but when, from any cause, the enamel becomes broken, the tooth decays, and the nerve is left bare, then the action of the air and irritating substances cause pain.

This pain is sometimes extremely severe, and in some cases is attended with a violent thrill at each pulsation of the heart and arteries, which is called "*jumping* toothache."

Fred.—How can the pulsation of the heart and arteries have any effect upon the teeth, as there are no blood-vessels in the teeth?

Father.—There are blood-vessels in the teeth. The nerve of each tooth is accompanied by a minute artery and vein,—the artery to supply nutriment to the tooth for its growth, and the vein to carry back the blood.

Fred.—Well, Father, I see I know but little about these things, and I am thankful to you for this instruction.

Father.—I must close the present interview with a remark upon the effect of the teeth on personal appearance. God designed the teeth for ornament as well as for utility. "The expression and general appearance of the face depends much upon the condition of the teeth. If they are perfectly regular, pure and clean, they contribute more to beauty than any of the other features; but if neglected, diseased, or incrusted with offensive accumulations, they excite in the beholder both pity and disgust." The influence which the teeth exercise over beauty is greater than all the other attractions of the countenance. This ornament is equally attractive in both sexes;—it distinguishes the elegant from the slovenly gentleman—it diffuses amiability over the countenance by softening the features.

But it is more especially to woman that fine teeth are necessary, since it is her destiny first to gratify the eye before she touches the soul, and captivates and enslaves the heart. The dark, black eye may be ever so piercing, the soft blue eye may melt with tenderness, the rose may blossom brightly upon a downy cheek, yet all charms lose their power if the teeth are defective.

Mary.—I suppose, Father, that you intended your last remarks for me.

Father.—It would be well, my daughter, for all misses to remember the lines of Moore,—

> " What pity, blooming girl,
> That lips so ready for a lover
> Should not beneath their ruby casket cover
> One tooth of pearl!
> But like a vase beside the church-yard stone
> Be doomed to blush o'er many a mouldering bone."

In a future conversation, I shall give instruction relative to the care and preservation of the teeth.

QUESTIONS ON CONVERSATION XVI.

PAGE 229.—Why has the infant no teeth? What will be required?

PAGE 230.—The Creator provided what? When do they make their appearance? What wonderful fact is here stated? How is losing the first teeth proof of God's goodness?

PAGE 231.—Repeat the substance of what is stated on this page?

PAGE 232.—What is provided underneath the first set of teeth? What do they become? What is said of the second set? What of the removal of the first teeth? Why is the providing of the second teeth more wonderful than the first.

PAGE 233.—What is the first set of teeth called? What is the number of the temporary teeth? What is the number of the permanent teeth? Why are there more of the second than the first teeth?

PAGE 234.—Give the names of the teeth. What is said of the last tooth on each side? Of what does each tooth consist? What is said of the root? What of the crown? Which have but one root? Which have two roots? Which have three? What is said of each division? Describe the incisors and their use. Give an example.

PAGE 235.—What is said of the molars? What more is said of the use of the different divisions? What is said of the composition of the teeth?

PAGE 236.—What do they require? Describe the covering of the teeth. What is said of its hardness? What is the internal part of the tooth called? Why do the teeth require different covering from the bones? How is this proved?

PAGE 237.—Have the first teeth roots? How then do they come out without roots? What is the object of the absorption of the root?

PAGE 238.—What is another important reason? What is said of such a condition?

PAGE 239.—What are the sockets called? What is said of the movements of the jaw? How is the first movement produced? How the lateral, or grinding?

PAGE 240.—What is said of these muscles? What in case of locked-jaw? What is in the centre of each tooth? What is said of this nerve? What produces "jumping toothache"? What accompanies the nerve of each tooth? For what purpose?

PAGE 241.—What is said of the effect of the teeth upon personal appearance? What is more especially necessary to women?

PAGE 242.—What would be well for misses to do?

CONVERSATION XVII.

THE SPECIAL SENSES.

THE SENSES THE ONLY MEDIUM OF OBTAINING KNOWLEDGE.

THE SENSE OF VISION.

THE ORGAN OF VISION—THE OPTIC NERVE—THE EYE—MUSCLES OF THE EYE—ORBITS OF THE EYE—PROTECTING ORGANS—HOW THE ACT OR FUNCTION OF SEEING IS PERFORMED—WHY WE SEE OBJECTS SINGLE AND NOT DOUBLE—THE FOCAL POINT OF VISION.

Father.—In previous conversations I have endeavored to show that all the organs and parts of our bodies perform *distinct functions*, each differing totally from all others.

Thus the bones give firmness and form to the body,—the muscles produce all its movements,—the nerves are the medium through which the mind controls the muscles,—the stomach digests the food,—the heart and blood-vessels circulate the blood,—the lacteals absorb the nutriment of the blood,—the secretory organs draw off its impurities, etc., etc.

Of the myriads of parts of which our bodies are composed, no two perform the same function. Some of these I have already described, but as yet I have said nothing, very definite, as to the mode and medium by which the mind holds connection, and communicates with the external world.

Fred.—I thought the mind was separate from the world.

Father.—I mean the ways or means by which the mind obtains knowledge.

Fred.—Do we not obtain knowledge by reading books, and by instruction at school?

Father.—Books and teachers are means of instruction, but without the senses they cannot be mediums of obtaining knowledge. What book or teacher could make you understand what light, darkness or color is, if you had no eyes;—or what pain is, if you had never felt it; or what sweet or bitter is, if you had never tasted either; or the fragrance of a rose, or the fetor of putrid matter, if you had never smelled them; or what music or thunder is, if you had never heard either?

Now all these are mediums by which we obtain knowledge. What can we know of any thing, in all the universe of God, unless we obtain the knowledge, either by seeing, hearing, feeling, tasting or smelling?

Fred.—What! are there only five ways, by which we obtain all that can be known on all subjects?

Father.—True, my son, and these are called the five senses; namely, the sense of Touch, Taste, Smell, Hearing, and Vision or Sight.

Office of the eye.

Frank.—Then God has made organs in our bodies for the purpose of giving us knowledge, has he?

Father.—He has, and most wonderful organs they are, too.

Mary.—I suppose they are called knowledge organs.

Father.—You can call them so my child, if you choose.

I shall now proceed to give you a brief description of the special senses, their organs and functions. I will commence with the organ of vision—

THE EYE.

The design and mechanism of the eye, is not surpassed, perhaps not equalled, by any other organ. The sense of sight is the most perfect, and contributes to our enjoyment and happiness more than any of the other senses.

The eye informs the mind of the color, form, volume, and position of the various objects by which we are surrounded. Through the eye, the mind perceives, and is delighted with the beauty of the rose, the lily, the landscape, the architectural proportions of edifices, and the splendors of the starry heavens. Vision guides us in our daily avocations, and guards from dangers that ever beset us. It is an essential means by which we obtain a knowledge of the sciences, and the arts of life. No one can estimate its value until deprived of its use.

246 ANATOMY AND PHYSIOLOGY

Optic nerve.

In form, the eye is a spheroid, though usually called a globe. It is nearly an inch in diameter. It consists of the *Op'-tic Nerve*, muscles, coats, humors, protecting organs, and several other parts, which I will briefly describe.

Fig. 39.

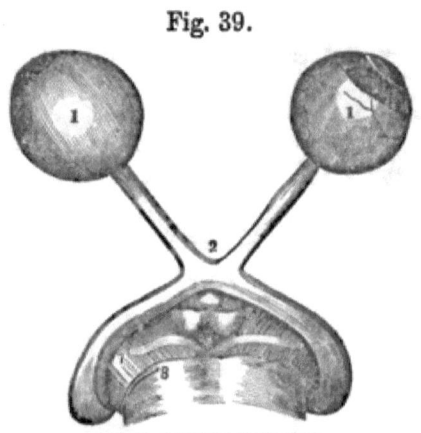

THE OPTIC NERVE.

Fig. 39. 1, 1, the globe of the eye; 2, the crossing of the optic nerve; 3, the origin of two pairs of cranial nerves.

The OPTIC NERVE arises from the central portion of the base of the brain, by two roots. As they proceed forward they diverge or separate from each other, and enter the globe of each eye at the back, and expand into a whitish membrane which constitutes the principal part of the *ret'-i-na*.

The BALL or GLOBE is a perfect optical instrument, constructed upon principles essentially the same as instruments formed and used by opticians. The sides of the globes are formed of coats or membranes, while

Names of the different parts of the eye.

Fig. 40.

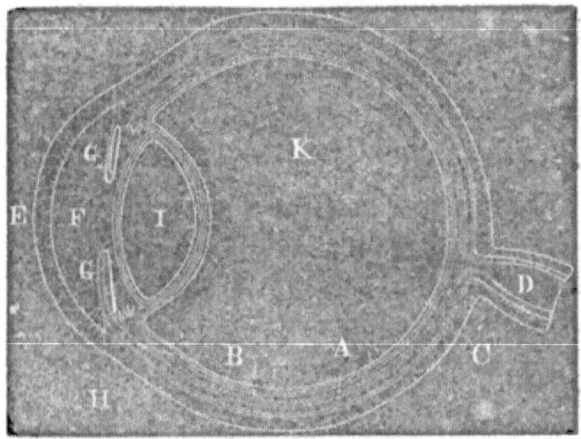

THE GLOBE OF THE EYE.

Fig. 40. E, a good view of the cornea fitted into the sclerotic coat; A, the choroid; C, the retina; K, the vitreous humor; D, the optic nerve; I, the lens; G, the iris, pointed on the back side with pigment; F, the aqueous humors.

the inside is filled with what is called refracting *Humors*.

There are three COATS or membranes,—1st. The *Scle-rot'-ic* and *Corn'-e-a*. 2d. *Cho'-roid*, *I'-rus*, and *Cil'-ia-ry* process. 3d. The *Ret'-i-na*.

There are also three HUMORS:—The *A'-que-ous* or watery, the *Crys'-tal-line* (lens) and the *Vit'-re-ous* or glassy.

The SCLEROTIC COAT is a dense, fibrous, and very firm membrane, which covers about four-fifths of the ball or globe of the eye. To this membrane, the muscles are attached, which move the eye. It is very white and is what is commonly called "the white of

Cornea, iris, pupil, retina.

the eye." The one-fifth of the eyeball not covered with the sclerotic coat, is the front part, and is covered by the CORNEA, which is a transparent layer, through which light is admitted. It is circular in form, and on the outside is convex, or oval, and on the inside is concave, like the glass of a watch,—or it is what is called convexo-concave. The cornea is firmly attached by its edges, to the sclerotic coat, and these two form the entire covering of the ball of the eye.

The CHOROID COAT is of a rich chocolate brown color on the outside, and black on the inside.

The IRIS forms a partition between the anterior or front, and the posterior or back chamber of the eye. Its color varies greatly in different persons. It has a circular opening called the *Pu'-pil*.

There are two layers of muscular fibres around the pupil, called radiating and circular. By the action of the radiating, the pupil is dilated or enlarged. By that of the circular, it is contracted.

Fred.—Why does the pupil relax and contract?
Father.—The object is very important, which I will explain at the right time.

The RETINA is formed of three layers,—the external, middle or nervous, and internal or vascular. The external is an extremely thin membrane. The middle is the expansion of the optic nerve. The vascular consists mostly of minute arteries and veins.

You will recollect that the RETINA forms the back part of the inside of the eye. Its importance in the function of seeing is great, as I shall show before we dismiss the subject.

Mary.—Father, you said that the globe or ball of the eye is filled with refracting humors. What are they?

Father.—They are fluids which crook, or bend the rays of light while passing through them.

Mary.—What! bend the rays of light!

Father.—I will explain it after I have finished the description of the parts of the eye.

There are six muscles of the eye, as I stated to you, I believe, in a previous conversation. These are attached to the bones of the orbit at one extremity, and at the other, by tendons, to the sclerotic coat. These muscles are so placed, and draw from such various points, that they move the eye in every conceivable direction, and place it in every possible position.

They also, sometimes, become permanently contracted at the outer or inner corners of the eye. If it be the muscles of the outer corner, the eye is turned out, producing what is called "wall eye." If it be the inner corner which is contracted, then it is turned in, causing what is called "cross eye."

Fred.—I have seen such cases; is there no remedy?

Father.—Yes, my son, such cases are easily cured by a skilful surgical operation, in partially cutting the contracted muscle, so as to allow the eye to turn back to its proper position.

Frank.—Will you now tell us about the protecting organs you before mentioned? Has God made organs on purpose to protect the eye?

Father.—He has;—and it is proof of the truth of the declaration of the Psalmist,—"His tender mercies are over all his works."

"The PROTECTING ORGANS are the *Or'-bits, Eye-brows, Eyelids,* and *Lach'-ry-mal Apparatus.*"

The ORBITS are deep cavities in the bones of the face, called sockets, wrought out, apparently, with the nicest mechanical skill.

There is a hole through the bone at the bottom of the orbit, for the passage of the optic nerve. The eye is imbedded in fatty matter which lines the socket, and serves as a soft cushion upon which the eye rests and performs all its movements without pain or irritation.

In youth and middle age, the socket is so filled with this adipose or fatty bedding, that the eye stands out prominently. Thus the figurative words of the Psalmist,—"Their eyes stand out with fatness," are literally true. In advanced age, and in protracted sickness, this matter is absorbed, and the eye sinks into the socket, and presents the appearance peculiar to the aged and diseased.

How clearly we discover, both wisdom and goodness in the location of the eye;—a deep cavity, with a bony rim, that affords a perfect protection to this delicate and important organ. So perfect is the protection, that if one in the darkness of night, should strike the face against the edge of an open door, the face might be injured above and below the eye, while the eye itself would remain untouched. Indeed, no part of the face is less liable to injury than the eye.

The EYEBROWS are projecting arches, forming the upper part of the orbits. They are covered with short thick hair, so laid, turning from the nose outward to the right and left, as to prevent the free perspiration flowing into the eyes. They also protect them by affording a shade from too intense light.

Another means of protecting the eye, is what you see every day, and yet, perhaps, have never given it a serious thought, I refer to the EYELIDS, which may be regarded as the *curtain of the eye.*

They are lined on the inside by a smooth membrane, called the *Con-junc-ti'-va*, which secretes the fluid that moistens and lubricates the eye.

The principal object and function of the eyelid seem to be, 1st. To moisten the eye during our waking hours, by passing frequently over the eyeball, also by its rapid motion to shield the eye from motes or any thing that might cause pain and injure its delicate structure. 2d. To cover it during the time of sleep, thus shielding it from harm during our unconscious hours.

This membrane—the eyelid, is both voluntary and involuntary in its action. We can open or close it at pleasure—its action is perfectly under the control of the will, and yet, most of its movements are made *entirely independent* of the will.

Who ever thinks to close his eyes on laying his head upon his pillow to sleep? Or who ever thinks to open his eyes on waking in the morning?

And yet they are sure to close on going to sleep, and to open on waking. Now here is an arrangement of the Creator, the wisdom and goodness of which are equalled only by its utility,—that muscles so perfectly voluntary, and under the control of the will, should act so perfectly *in*voluntary and independent of the will, simply because our safety, convenience and comfort require it. Muscles which we can move as we please, yet when we forget or neglect to do so, they move themselves, and always in the right direction. When weary nature indicates the need of sleep, whether by day or night, the curtain drops and shuts out both the light and the world; and we sink into quiet repose; and on waking, the curtain is raised and the light again greets and illuminates the eye,—and all this without the least care of ours.

Now as we seldom or never think to close the eyes on going to sleep,—did not the eyelids close involuntarily, the eyes would be glaring open, dry and glazed, exposed to insects and dust, during the large portion of life which we spend in sleep. Now how long do you suppose we should have eyes, were they not protected by the involuntary action of the lids?

Fred.—Not long, certainly. How numerous the proofs of God's kind care for our safety and comfort!

Fig. 41.

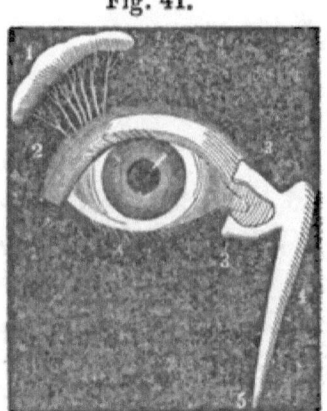

THE LACHRYMAL GLAND.

Fig. 41. 1, the lachrymal gland; 2, the ducts leading to the upper eyelid; 3, 8, the tear points; 4, the nasal sac; 5, the termination of the nasal duct.

Father.—The LACHRYMAL GLANDS, which secrete the tears, are situated at the upper and outer angle of the orbit. The gland is oval shaped, and about three quarters of an inch in length.

The lachrymal secretions—the tears.

Connected with this gland are several small ducts which open upon the upper lid, and pour the tears upon the eye, which, from this cause, is always kept moist. How wonderful is this arrangement!

The lachrymal secretion taking place at the upper part of the eye, so as to flow down over the ball; and the lid, involuntarily and unconsciously to ourselves, moving up and down over the eye almost every waking moment of life, for the purpose of spreading these secretions over the eye to keep it moist.

Fred.—What would be the effect, Father, if the quantity secreted should be more than is needed to moisten the eye? I suppose in that case, the tears would flow over the face.

Father.—Such would be the result, had no provision been made to meet such an emergency. But this necessity is provided for. There is a canal, called the NASAL DUCT, which leads from each eye to the cavity in the nose, or nostril, by which any excess of the lachrymal secretion, or tears, is carried off. (See Fig. 41—4, page 253).

There is one thing worthy of note, in regard to the lachrymal secretion, which is, that when the eye becomes irritated by means of dust, motes, or any other substance, these glands pour out a large quantity of tears for the purpose of washing away the irritating cause.

I have now briefly described the principal parts of

the eye concerned in the act of seeing. Do you understand any better, from what I have said, in what way, or how the act or function of seeing is performed?

Fred.—I do not. I can only say that I see with my eyes!

Father.—But *how* do you see?

Fred.—I look straight at a thing. The sight goes from my eyes to the object I look at.

Father—Is that the best explanation you can give? What do you say Frank? Can you give a better answer?

Frank.—No, Father. I thought, as Frederick said, the sight goes from the eye to the object we see.

Father.—What do you say about it, Mary?

Mary.—I thought we see in the way that Frederick and Frank has said we do.

Father.—My children, the *reverse* of what you have said is true, if I understand what you mean. Instead of the sight going from the eye to the object seen, it is the object seen that comes to the eye, and not only so but goes into it, and remains at the farthest or back part of it.

Fred.—Do you mean to say that when I see a man, he comes into my eye?

Father.—Not exactly that; I mean that when you see a man, his exact image or picture, is formed on the *retina*, and is conveyed to the brain and mind by the optic nerve.

And it is so with each and every object we behold, whether great or small, and this produces the sense of vision.

Fred.—Well, Father, that is certainly a new idea. Does then the picture formed on the retina contain the exact shades of color, as well as the form of every thing we see?

Father.—It does. Color, form, size, every thing in the picture, is perfect.

Frank.—You surprise me, Father! Have I always had pictures formed in my eyes,—pictures of the beautiful flowers of the garden and field,—the clouds and starry heavens, and yet never knew it?

Father.—Such is the fact, my son.

Frank.—I should be pleased to know how the image or picture is formed on the retina.

Father.—To understand the principles by which the image is formed on the retina in the act of seeing, involves a knowledge of the science of *Optics*. To be acquainted with the structure of the eye, is not sufficient to give a knowledge of the theory of vision. To comprehend the theory of vision, you must be familiar with the properties of *light*. And this requires a more extensive knowledge of the science of natural Philosophy than you have yet obtained. I shall, therefore, defer a *minute* explanation of the *theory* of vision, until you are better prepared to understand it. I will, however, state some general propositions in regard to it.

The principles of vision.

Rays of light emanate from every object we see. These, as they enter the eye through the pupil, are refracted by the humors of which I have spoken, and thereby are so arranged as to form upon the retina, the exact image of the object from which they proceed.

The principle by which the picture is formed upon the retina, is essentially, if not precisely the same, as that by which the artist produces his picture upon the prepared glass placed in the instrument called the camera,—the glass taking the place in the camera that the retina does in the eye, and the *lenses* in the camera, like the humors in the eye, refracting and converging the rays of light to form the picture.

Mary.—Do you mean, Father, Ambrotype pictures and the like?

Father.—Yes, my child, I mean all kinds of pictures that are obtained by the use of the camera, or the *Cam'-e-ra Ob-scu'-ra*, as it may be more properly termed.

Mary.—What is that, Father?

Father.—It is an optical machine or instrument used in a darkened room, for throwing the image of an external object upon a plain surface. Its principle of action, as I have before said, is essentially like that of the eye in forming pictures. There is this difference, it is true, showing how much more perfect are the works of God than those of man. The picture of persons, or landscapes, or any object obtained by the artist by the use of the camera, is formed upon glass previously prepared with chemical substances, to absorb and retain the rays of light thrown upon it, by which the picture is formed;—whereas the retina is always ready to receive the picture.

Mary.—Yes, Father, but the picture made by the artist upon the glass, will last a great many years, and be just as bright as it was at first. But the picture made upon the retina is gone as quick as the person or object is out of sight.

Father.—That is true, my child, and it is additional proof of God's wisdom and goodness.

Mary.—How so? What would the artist's picture be worth, if it was as easily and quickly defaced and destroyed as the picture on the retina?

Father.—It would be worthless, it is true,—but the picture of the artist was *designed* to be durable—its value depending upon its permanency. Now let us suppose the picture formed upon the retina to be equally lasting, what would be the result? Suppose you look out upon the fields and forest spread out before you,—the picture of a beautiful landscape would be formed upon the retina, *and there it would remain.* You would ever after see a beautiful landscape, but could never see any thing else. The retina would all be occupied by the landscape, and no other picture could be formed upon it. Father, Mother, Brother, Sister, Friends, books, you could never again see. The goodness of the Creator is most clearly seen, in the endowment of the retina with the ability instantly to receive the picture, and to part with it as quickly, thus giving room for others, and if need be, totally different ones, and thus thousands may be formed in a short time, each as perfect as if there had been but one.

Mary.—It is wonderful!

Frank.—Father, you promised to explain the object and use of the radiating and circular fibres around the pupil, by which it is dilated and contracted.

Father.—The pupil, you recollect, is a circular opening in the iris, through which light is admitted and falls upon the retina. Now the retina, being the expansion of the optic nerve, is very sensible to the quantity of light thrown upon it.

To illustrate.—If we pass suddenly from darkness into a brilliant light, it dazzles the eye, and causes severe pain, and the pupil contracts and shuts out a part of the light. If we pass from a bright light into a dark room, at first we can see no objects around us, but in a short time the pupil expands and lets in more light, and we at length are able to discover objects that we could not see before.

In a word, when the light is so intense as to cause pain, the pupil contracts for the purpose of shutting some of it out. When there is not light enough, it then expands to let in more. And this action, so perfect and so important, is entirely involuntary. It takes place without a care or thought of ours. And, what is more astonishing, it is regulated by the light itself,—too much causing the pupil to contract, and too little making it expand. In short, it is a perfectly self-regulating apparatus, controlled entirely by the intensity of light present at any given time.

Now who cannot see, that when the Creator designed and formed the eye, all these circumstances were taken into consideration, and this necessity, so important to our comfort, duly provided for by the formation of the muscular fibres whose office it is to control the size of the pupil, and thereby the quantity of light admitted to the eye.

Had the Creator been indifferent to our comfort, can we suppose we should find such an arrangement, in the eye? But for this, the size of the pupil would always be the same, and during every clear, sunny day, the eye would be dazzled and pained, if not blinded, by the intensity of the light,—and in a cloudy day, one would be in almost total darkness. Should we not say with the Psalmist,—"All thy works praise thee."

Frank.—I think we should. I did not suppose so many proofs of God's goodness could be drawn from this one sense.

Father.—I have not spoken of them all, and shall not. But there are one or two others, which I must name before we close our present interview.

Focus of sight defined.

In the act of seeing, there is an image of the object seen, formed upon the retina of *each* eye, still we see but one object.

Fred.—Why do we not see *two* objects instead of one?

Father.—It is because the two pictures, being exactly alike, convey but one sensation or impression to the mind. The rays of light, which form the picture upon each retina, proceed from one object, which is what is called the *Fo'-cus* of sight,—that is, it is the point where the rays of light meet. And every object we see, whether great or small, near or distant, constitutes the focal point of vision. And in order to have perfect vision, each eye must be turned in such a position as to allow the rays of light passing from the focal point through the pupil, to fall fairly upon the retina. Otherwise, two images or objects, instead of one will be seen.

Now in the light of this fact, we may see what perfection there is in the formation and action of the eyes. In moving them from one object to another, in order to perfect vision, the movements of both must be *exactly alike*. Should one move a little farther, or not quite so far as the other, vision would either be destroyed, or we should see things double. You will perceive, by moving the eyes in every direction, and to the greatest possible extent, that vision is as perfect with the eyes turned in one position as another, showing that the movement of each eye-ball perfectly accords with that of the other, and, that although each is moved by a distinct set of muscles, they are as perfectly alike, as if both were acted upon or moved by a single muscle. And the exactness of movement of each is not accidental, it is a matter of *absolute necessity*, for vision without it would be impossible.

I will just mention one thing more which must close this long conversation. Although there can be but one focal point, and consequently but one object seen distinctly at a time, yet while you are looking directly at that object, you can see many others,—*less distinctly*, at the same time. For example :—Let Frank and Mary stand in opposite corners of this room, and Frederick stand midway between them. While I look directly at Frederick, and he is the focal point of my vision, at the same time I see Frank and Mary with sufficient distinctness to notice any movements they may make. I look into the street, and while I place my eyes upon an object, that is, while one, say a horse, becomes the focal point of my vision, I can, at the same time, see many other horses, and men;—less distinctly, it is true, yet sufficiently so to see what they are doing.

Frank.—What particular benefit is derived from it?

Father.—Very great, my son, and that you may know it, let us suppose that while one was looking at something, every thing else was shut out from view, the same as if you were looking through a tube. In that way how impossible it would be to avoid the dangers with which we are always beset, especially in the streets of large cities. Indeed it would be *dangerous* to go into them, for, seeing but one object at a time, we should be ignorant of any danger, and consequently could not avoid it.

A frantic horse might be coming upon us, but not knowing it, even if we heard him, we should not know which way to escape, or if we attempted it, we might run into greater danger. But with the present capacity of the eye, one can be looking upon a single object, and at the same time see all that is going on within the range of his vision, and thus avoid danger. But I will not now detain you longer. By due reflection you will see how very valuable are the endowments of the eye.

Fred.—How large is the picture that is formed upon the retina?

Father.—That depends upon the size of the object we look at, and its distance from us.

The picture of a landscape of a hundred acres, with its hills, houses, trees, etc., would not be larger than a dime;—and pictures of other things would be in proportion;—by which you see what extreme perfection there must be in the picture to show every form and feature of the object seen.

Fred.—Father, you promised to tell us how the refracting humors bend the rays of light.

Father.—Light is refracted, or turned from its course, when it passes obliquely from one medium to another of different density, as from air to water, or from water to air. You have doubtless noticed that the part of a boatman's oar which is under the water, appears to be bent? This is because the rays of light, in passing from the oar in the water to the eyes, are bent or crooked. In the same way, the humors of the eye turn or bend the rays of light which enter it, so as to throw them upon the retina to form the picture.

Fred.—I do not know, Father, how to express my gratitude for the instruction you have given us.

Father.—All I wish in return is, that you will make a good use of the knowledge you obtain.

QUESTIONS ON CONVERSATION XVII.

PAGE 243.—What has been shown in previous conversations? Give examples. What is said of the myriads of parts of which our bodies are composed?

PAGE 244.—What is said in regard to the ways or means by which the mind obtains knowledge? What is the number of ways? What are they called?

PAGE 245.—What is said of the design and mechanism of the eye?— What is said of this sense? Of what does the eye inform the mind? What does the mind perceive through the eye? What is said of vision? Of what is it the principal means? When can its value be appreciated?

PAGE 246.—What is the form of the eye? What is its diameter? Of what does it consist? Describe the optic nerve. Into what does it expand? What is said of the ball or globe?

PAGE 247.—How many coats or membranes are there? What are their names? How many humors are there? Give their names. Describe the sclerotic coat? What does it cover? What is it commonly called?

PAGE 248.—What is covered by the cornea? What is admitted through the cornea? What is the form called? To what is the cornea attached? What forms a covering for the eye? What is the color of the choroid coat? What does the iris form? What is said of its color? What is its opening called? What are around the pupil? What are they called? Describe their action. What is important? How is the retina formed? Describe their layers.

PAGE 249.—What is said of the retina? What are refracting humors? What is the number of muscles of the eye? To what are they attached at each extremity? How are they placed, and how do they draw? What is the result? What is the effect of a permanent contraction of these muscles?

PAGE 250.—What is said of a cure of such cases? What is said about the protecting organs? Of what is it proof? What are the protecting organs? Describe the orbits. What is there at the bottom of the orbits? In what is the eye imbedded? What is its use? What is said of the socket in youth and middle-age? What in advanced age and in sickness?

PAGE 251.—What is said of the location of the eye? Describe the eyebrows and their use. What is said of the eyelids? With what are they lined, and what is its function? What is the object and function of the eyelid?

ANATOMY AND PHYSIOLOGY.—CONVERSATION XVII.

PAGE 252.—What is said of its action? What is said of this arrangement.

PAGE 253.—What would be the result, did not the eyelids close involuntarily? What are numerous? Describe the lachrymal glands.

PAGE 254.—What are connected with this gland? What is their function? What is said of the nasal duct? What is worthy of note?

PAGE 255.—How is the act or function of seeing performed? What is formed on the retina? Where is it conveyed? By what?

PAGE 256.—What does this produce? What does the picture contain? Does a knowledge of the structure of the eye explain the theory of vision? What does a knowledge of the principles of a vision involve?— With what must you be familiar to comprehend the theory of vision? What does this require? What will therefore be defered?

PAGE 257.—What general propositions in regard to the theory of vision are stated? The principle is the same as what? What is a camera obscura? What is the difference here spoken of?

PAGE 258.—What is said on this page?

PAGE 259.—In what is the goodness of the Creator seen? Explain the object and use of the radiating and circular fibers around the pupil? Give an illustration.

PAGE 260.—What is said of this action? By what is it regulated? What more is said of it? What was taken into consideration when the Creator designed and formed the eye? What more is said about this arrangement? What should we say with the Psalmist?

PAGE 261.—What is done in the act of seeing? Why then do we not see two objects instead of one? What is said of the rays of light? What is the focal point of vision? Every object seen constitutes what? To have perfect vision what must be done? Otherwise what? What may we see in the light of this fact? What more is said of the movement of the eyes?

PAGE 263.—What is said about seeing more than one object at a time? Give examples. What benefit is derived from this? Give examples.— What does the present capacity of the eye enable one to do? What is said about the size of the picture formed upon the retina? What do we see by this?

PAGE 264.—How do refracting humors bend the rays of light? Give an illustration. By what is this caused? What do the humors of the eye do in the same way?

CONVERSATION XVIII.

SENSE OF HEARING.

ITS IMPORTANCE—THE ORGAN OF HEARING—A DESCRIPTION OF THE EAR—THE AUDITORY NERVE—THE FUNCTION OF HEARING, HOW PERFORMED, OR HOW SOUND IS PRODUCED—CAN BE NO SOUND WHERE THERE IS NO EAR—HOW AN ECHO IS PRODUCED—THE PRINCIPLE OF ACTION OF THE SPEAKING TUBE, AND EAR-TRUMPET—THE CAPACITY OF THE EAR—THE POWER OF SOUND TO CONTROL THE FEELINGS OR EMOTIONS OF THE MIND.

Father.—I intend, at our present interview, to give you a description of the sense of hearing, which, in its importance, is next to that of seeing. By this sense we derive a great amount of knowledge from listening to lectures, sermons, debates, instruction from teachers, and in social converse.

This sense affords us *pleasure* by listening to the deep tones of the organ, the sweet strains of the human voice, and the warbling of the feathered songsters.

The organ of hearing is the EAR, which, in its structure, is one of the most complicated in the human body. Its parts are numerous, the function of some of which is not well understood. I shall give you only a general description of the parts and functions—such as will best subserve our purpose, believing that a *minute* description would rather confuse than aid your understanding.

Fig. 42.

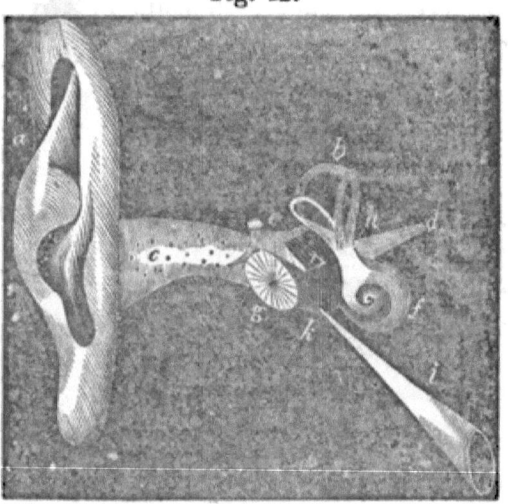

THE EAR.

Fig. 42. a, is the pavilion or rim of the ear; c, the auditory canal; g, the membrana tympani; k, the tympanum; e, the bones of the ear; b, the semicircular canals; f, the cochlea; h, the vestibule; i, the eustachian tube; d, auditory nerve.

The EAR may be divided into three parts,—1st. The *External Ear*. 2d. The *Tym'-pan-um*, or middle Ear. 3d. The *La'-by-rinth*, or internal Ear.

The EXTERNAL EAR is the *Pin'-na*, (pavilion or rim) and the *au'-di-to-ry ca-nal!*

The PINNA is a broad plate, placed around the entrance of the auditory canal. It is formed of cartilage, and presents many ridges and furrows.

The parts of the ear.

The AUDITORY CANAL is about an inch in length, and extends inward to what is called the *Tym'-pan-i*, (drum of the ear). Short, stiff hairs are formed in this canal, which stretch across it, and were evidently designed to prevent the entrance of insects. There are also, secretory glands or follicles in this canal, which secrete a very bitter substance called ear-wax. This, also, was doubtless designed to prevent foreign ingress.

The TYMPANI is a thin membrane, of an oval shape, and about three-eighths of an inch in diameter. It is stretched across, and connected with, the external canal near its termination, and separates the external from the middle ear.

The MIDDLE EAR, or TYMPANUM, is an irregular, bony cavity, filled with air. It contains four small, curiously shaped, movable bones, which form a connection between the external and internal ear. There is a canal called the *Eu-sta'-chi-an Tube*, which forms a connection between the Tympanum and the *Phar'-ynx*, or back part of the mouth, the object of which is to admit air into the tympanum, which, by pressing equally upon both sides, keeps the drum in a proper state of tension. Without this air no sound can be produced.

The LABYRINTH is situated between the tympanum

The auditory nerve.

Fig. 43.

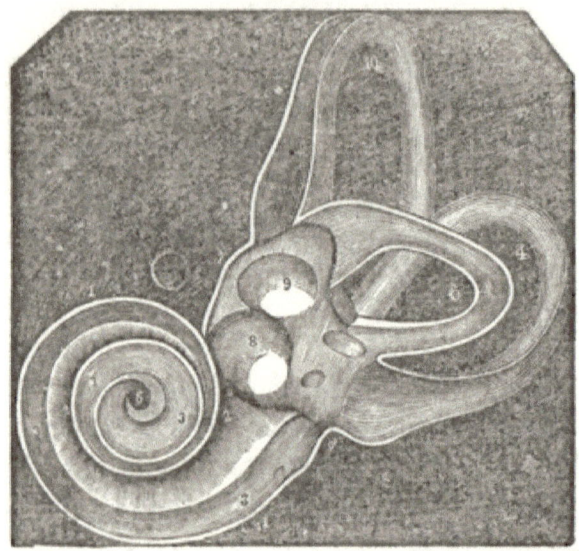

A VIEW OF THE LABYRINTH LAID OPEN, AND HIGHLY MAGNIFIED.

Fig. 43. 1, 1, the cochlea; 2, 3, the channels that wind around the central point (5); 7, 7, the vestibule; 8. the foramen rotundum; 9, the fenestra ovalis; 4, 6, 10, the semi-circular canals.

and the *Aud'-it-o-ry Nerve.* It consists of bony canals filled with a fluid.

Fred.—Father, you say, that the middle ear is filled with air, and the labyrinth or internal ear with a fluid, but you did not say what separates them.

Father.—They are separated by a membrane which may be regarded as a second drum, the use of which I will explain hereafter.

The AUDITORY NERVE proceeds from the brain, enters the temporal bone, as it is called, and divides into two branches at the bottom of the internal ear. These branches radiate in all directions, upon the inner surface, in a manner somewhat resembling the optic nerve to form the retina.

I have now described, very briefly, the most important parts of the mechanism of the organ of hearing.

Now, my children, have you any definite idea how the function of hearing is performed, or how sound is produced.

Fred.—My ideas are confused. I suppose the sound finds its way in and gets to the auditory nerve, although the passages are so crooked, and filled up with so many drums and other obstructions, that I should think the ear would be perfectly deaf.

Father.—These "drums and other obstructions," as you call them, are parts of the organ of hearing, and so far from forming obstructions, there could be no sound produced without them.

The function of hearing, or sound, as we call it, is produced by vibrations of the air falling or striking upon the tympani or drum of the ear.

Frank.—Father, what causes vibrations of the air?

Father.—Any thing in violent motion. When a bell rings, there is a trembling or vibration of the bell, which produces vibrations of the air that surrounds it. When you strike a key of the piano, or draw a bow across the string of a violin, the string quivers or vibrates, and this communicates the same to the air.

And when we speak or sing, the vibrations of the vocal chords produce the sounds. Thunder causes so much vibration of the air, as, in some instances, to shake or jar the ground. The discharge of a cannon produces vibrations that extend for miles around.

Frank.—How does that produce sound?

Father.—This motion of the air falls upon, and causes a vibration or quivering of the drum of the ear, and this produces a vibration of the four little bones, and also of the air in the middle ear, and this motion is conveyed to the internal drum and the fluid which fills the internal ear, and this conveys the vibratory motion to the auditory nerve, which carries it to the brain and mind, and thus produces the sense of hearing. Thus you see sound is produced wholly of vibrations;—1st. The vibration of some body or substance. 2d. Of the air. 3d. Of the drum of the ear. 4th. Of the little bones and the air of the middle ear. 5th. Of the second drum. 6th. The vibrations of the fluid which fills the internal ear, and conveys them to the auditory nerve, by which the sensation is conveyed to the mind. Every sound we hear, requires all the vibrations I have named. Let the sounds be ever so short, or follow each other ever so rapidly, still, the vibrations producing each sound, are perfectly distinct,—they are never jumbled together, although they may chase each other, so to speak, with great rapidity.

Fred.—But, Father, sometimes there is a continuous sound for a long time, how do vibrations produce that?

Father.—When there is a continuous sound, there is a continuous vibration which produces it.

Fred.—I always thought when I heard a bell ring, the sound came right from the bell and went directly into my ear. I never supposed that so much shaking was necessary before I could hear any thing. I should like to know how vibrations can produce so many *different* sounds. Why have we at one time a low or weak, and at another, a loud sound? Why is one soft and pleasant to the ear, like the strains of the Æolian harp, and another, harsh and grating, like the touches of a file or the screeches of an owl?

Father.—The difference in sound is produced by *degrees of violence* in the vibrations which cause them,—some being mild as the ripples upon the placid lake, only moved by the gentle zephyr, giving sounds which are pleasant and agreeable,—while the harsh and disagreeable are produced by the violent agitations of the air, falling upon the ear like the dashing of furious waves upon a rocky shore.

There is one interesting fact in regard to the capacity of the ear which I must not fail to name here. It is that the ear is capable of hearing, with perfect distinctness, several different sounds at the same time.

Frank.—What benefit is there in that?

Father.—Very much, my son.

Frank.—Will you please to tell us what it is?

Father.—I will give you one instance out of many. Could we hear but one sound at a time, all the concords and harmony in music would be lost. To illustrate :—In an octave, there are several different sounds, which produce in music what is called *concord*. These are the first, or key note, the third, the fifth, and the eighth or octave. Now when these are all given together, either by different voices, or by the organ or piano, the ear hears each of them with equal distinctness. But we should have forever remained ignorant of all these sweet and harmonious sounds, were the ear capable of hearing but one at a time.

Fred.—I would like to know how vibrations can produce different sounds at the same time. You told us that those which produce each sound are never jumbled together.

Father.—You recollect I told you that the different sounds were produced by the different degrees of violence, or rapidity of the vibrations of the air. We might call them coarse and fine, the coarse producing the low, and the fine the high sounds. And while the coarse are acting upon the ear, causing the lower, the fine may, at the same time, produce the higher sounds.

Mary.—Well, Father, I am thankful that God has

given us ears,—I am so delighted with music that I should feel sad indeed, if I had no ears, for my piano would then be of no use, while now it is a great gratification to me.

Father.—My daughter, your piano would make no music, if you had no ears.

Mary.—Would make no music!—how could that be?

Father.—I mean what I say; that your piano would make no music if you had no ears to hear it.

Mary.—If I should become deaf, would my piano feel so badly about it that the keys would refuse to give their accustomed sound?

Fred.—What is the meaning of all this?

Father.—I mean simply, that there can be no sound where there is no ear to hear.

Fred.—Do you mean that the discharge of a cannon would make no sound, or that there would be no thunder if there were no ears to hear?

Father.—That is what I mean.

Fred.—Then it is the ears that produce the sound, is it?

Father.—No, it is no more the ears that produce sound, than it is the eyes that cause sight. Could the rays of light produce vision if there were no eyes?

Fred.—O, I think I understand now. You mean that, as it is the vibrations of the air upon the drum of the ear that causes sound, there can be no sound where there is no ear for the vibrations to act upon.

Father.—That is correct, my son.

Fred.—There is one thing in reference to sound that I could never understand, which is, why we can hear a person speak so much more plainly in a house, or a church, than in the open air.

Father.—It is for the reason that, in the open air the vibrations spread off in every direction, but in a house they are shut in by the walls of the room, and are reflected from one to another, consequently are stronger.

Mary.—Father, can you tell us what produces an echo?

Father.—That is very easily explained. When we speak, or halloo in the open air, the vibrations go off if there is nothing to obstruct them. But if they come against a house or any thing else, they are thrown back, or rebound, and strike the drum of the ear, causing an echo.

Mary.—I wish to ask why we can hear so plainly through a long speaking tube?

Father.—The vibrations are shut in and rebound from one side of the tube to the other, and as they cannot escape, they go directly through the tube. Persons miles apart, may converse in this way.

Mary.—Of what use is the rim of the ear?

Father.—The office of the pinna or rim of the ear, is to catch and direct the vibration into the ear,—for the more numerous these are that strike upon the drum, the more distinct will be the hearing. Some persons who are very deaf, use what is called an ear trumpet. This instrument is made thus :—One end is large, like a tunnel or a common trumpet, while the other is small enough to go into the external ear. In using it, the little end is inserted, and the large one, turned towards the person speaking, catches the vibrations and turns them into the ear.

Fred.—You remarked that we are capable of hearing a great variety of sounds. Can you tell us how many?

Father.—I cannot. The capacity of the ear in that respect, is beyond computation or even conception.

Fred.—You speak as though we are capable of hearing millions of different sounds.

Father.—Millions but poorly express the capacity of the ear.

Fred.—I am surprised at such a statement.

Father.—Do you suppose there can be found two persons, the sound of whose voices is so exactly alike that no shade of difference can be detected?

Fred.—I do not know. What do you think, Father?

Father.—I believe we may regard it as certain, that no two persons can be found on earth, whose voices are precisely alike. Were it possible for us to hear every person on earth speak, we should, I think, find a difference in each voice.

Fred.—I do not recollect ever having heard two persons speak alike.

Capacity of the ear.

Father.—I doubt not that it is so. Consequently, each person on earth, in speaking, makes a sound different from all others, so that the number of different sounds is equal to the number of persons living, or to the population of the earth,—and can you tell me what that is?

Fred.—About one thousand millions.

Father.—Consequently, the ear is capable of distinguishing one thousand million sounds. But this is not all; for there are many thousand species of birds, no two of which sing alike. And in each species are many million individual birds, and no two of these sing alike,—that is, each produces notes differing from all the others. Indeed, the same may be said of all animate nature. There is no living creature, either man, beast, bird or insect, capable of making a sound, which does not make it different from all others. Now, do you not see, that the number of sounds we are able to distinguish, far exceeds all human computation, and is absolutely incalculable?

The same thing is true of the different musical instruments. Let the same note be struck by every kind or sort ever invented, and although there may be perfect unity and concord of sound, yet that produced by each instrument, will be very different from that of any other. A perfect unity and concord may exist, yet a total dissimilarity of sound.

Fred.—I do not understand. You say a unity of sound, and yet a variety also.

Father.—I will illustrate. Let the same note be struck by the flute, violin, organ, and piano, and although the sound is a perfect chord, yet the same sound given by each, is so totally different from all the others, that it may be readily known which instrument produced it.

Fred.—I am sure that is true, for I can easily distinguish by the sound, the different musical instruments with which I am acquainted.

Father.—Well, now multiply all the tones and semitones that each of these instruments are capable of giving, by the number ever invented, and what a vast amount will be the sum total.

Now the facts which I have named, show an accumulation of numbers which set all computation at defiance, and show that the capacity with which the Creator has endowed the ear, borders upon infinity.

Fred.—It is most wonderful that this variety of sounds is caused by the different vibrations of the air.

Father.—Another important faculty of the ear is that it possesses the power or ability to determine the direction from which sound proceeds; and also, in a good degree, to form a judgment of the distance from which it comes. This is of great practical importance in the every day affairs of life. It aids to escape danger. It gives us the direction and the locality from which the cry for aid, as of a drowning man, comes.

Emotions produced by sound.

It informs the General as to the location of the cannonading he may hear. It directs the herdsman, by the tinkling of the bell, to his flocks and herds.

Sound has the power also, of controlling the feelings or emotions of the mind, and thereby affords pleasure, or causes pain.

"Music," it is said, "will tame the savage." Whether this be so or not, it is certain that it has a controlling influence over our feelings, inclining us to be lively or sad, grave or gay, according as it is melancholy or exhilarating.

Another peculiarity of the sense of hearing is, that the emotions and feelings are expressed by the tones of the voice. We readily understand those which indicate kindness, pity, rage, sarcasm, or contempt, even if they are expressed in a language of which we are ignorant. This knowledge is common to man, and children, at an early age, learn to interpret it; even brutes seem to understand it.

The tones of the voice not only have the power of *conveying* emotions, but also of awakening similar feelings in the mind of the hearer. A shout of victory from one part of an army sends a thrill throughout the whole, and nerves every arm. A shriek of terror will convulse a whole assembly. Expressions of anger tend to beget the same in the person to whom they are addressed, and those of affection give rise to similar emotions.

Reflections.

Thus you see, my children, in a limited degree, the inestimable blessing God has conferred upon us, by endowing us with the sense of hearing. Without it we should forever have remained ignorant of all that I have named, and all the knowledge possible to be obtained by this sense.

Frank.—We feel very thankful for the instruction you have given us, and hope we shall ever be grateful to our heavenly Father, that we have not, like thousands of our fellow beings, been deprived of this valuable sense.

Father.—We can never be too thankful, my children, for the perfect use of all the faculties of our bodies. And now, as a proper subject for reflection at the close of our present interview, let us recall the words of the inspired book,—

"He that planted (made) the ear, shall he not hear?"
"He that formed the eye, shall he not see?"

QUESTIONS ON CONVERSATION XVIII.

PAGE 265.—What is the subject of the present conversation? What is said of this sense? How does it afford us pleasure? What is said of the ear? What of its parts and their functions?

PAGE 266.—Into how many parts may the ear be divided? What are they? What is the external ear? What is the pinna?

PAGE 267.—What are formed in this canal, and what is their use? What is said of glands in this canal, and what do they secrete? Describe the tympani. With what is it connected? What does it separate? Describe the middle ear or tympanum. What does it contain? What canal is named? Between what does it form a connection? What is the object of this connection? Describe the labyrinth?

PAGE 268.—Of what does it consist? What separates the middle from the internal ear? Describe the auditory nerve.

PAGE 269.—What is said of these branches? How is sound produced? What causes vibrations of the air? Give examples.

PAGE 270. How do vibrations produce sound? Name the vibrations that take place in the production of sound. Every sound requires what? What more is said of the vibrations?

PAGE 271.—Explain how vibrations produce different sounds, such as low or weak, and loud,—soft and pleasant, harsh and grating? What interesting fact is stated?

PAGE 272.—What benefit is there in this? Give illustrations. How can vibrations produce different sounds at the same time?

PAGE 273.—Can there be any sound where there is no ear? Give an explanation of the fact?

PAGE 274.—Why can we hear more plainly in a house than in the open air? In what way is an echo produced? Explain the principles of a speaking tube. What is the office of the rim of the ear?

PAGE 275.—What do deaf persons use? What is its form? How is it used? What does it do? What is said of the capacity of the ear? Give illustrations.

What more is said on pages 276, and 277? What other faculty of the ear is named? What is said of its importance?

PAGE 278.—What has sound the power to do? To what does it incline us? What is another peculiarity of the sense of hearing? What do we readily understand? What is said of this knowledge? What power have the tones of the voice? Give examples?

PAGE 279.—What is conferred upon us by the sense of hearing? What if we were without it? For what can we never be too thankful? What is named as a proper subject for reflection?

CONVERSATION XIX.

THE SENSE OF SMELL.

THE NOSE THE ORGAN OF SMELL—ITS STRUCTURE—THE OLFACTORY NERVES—THE SENSE OF SMELL, HOW PRODUCED—OBJECTIONS TO THE THEORY CONSIDERED—THE SENSE OF SMELL A SOURCE OF PLEASURE—GOD'S GOODNESS IN PROVIDING IT.

THE SENSE OF TASTE.

THE TONGUE THE PRINCIPAL ORGAN OF TASTE—CAPACITIES OF THE TONGUE—THE TONGUE A DOUBLE ORGAN—THE THEORY OF TASTE—THE USE OF TASTE—PERVERTED TASTE—THE PLEASURES OF TASTE—INJURIOUS HABITS FORMED ONLY BY THE USE OF NARCOTICS—GOODNESS OF THE CREATOR IN PROVIDING THE SENSE OF TASTE.

THE SENSE OF TOUCH.

RESIDES IN THE SENSITIVE NERVES—THE VALUE OF THE SENSE—THE PLEASURES IT IS CAPABLE OF AFFORDING—GENERAL REMARKS UPON THE FIVE SPECIAL SENSES.

Father.—I have given you, my children, at our two last conversations, a brief description of the sense of vision or seeing, and the sense of hearing. At our present interview I shall speak of the three remaining senses—smell, taste and touch.

THE SENSE OF SMELL.

The organ of the sense of smell is the *Nose*, and

Organs and nerves of smell.

the sense is located in the air-passages of the nose. The organ is formed of Bones, Fibrous Cartilage, and the integuments which cover them. It is divided into two compartments called the nasal cavities or nostrils, extending to the pharynx or back part of the mouth. The nostrils, like all internal cavities of the body, are lined with mucous membrane.

The nerves, whose special office it is to produce the sense of smell, are called the olfactory nerves, and are spread upon the mucous membrane of the nose. The *acuteness* of the sense of smell depends upon the extent of the mucous membrane. In man it is considerable, but in animals which seek their prey by scent, the extent is far greater.

Frank.—I comprehend what you have said, Father, but I do not understand how the sense of smell is produced. I think I know now very well, since you have explained it so clearly,—with respect to the sense of seeing and hearing,—the sense of seeing by a picture of the object seen, being formed upon the retina, and hearing by vibrations of the air falling upon the drum of the ear. Now if you can explain as clearly with regard to the sense of smell I shall be glad.

Father.—The sensation of smell is produced in the following manner.

Extremely minute particles,—or what is called effluvia, are given off by odorous bodies, and float in the air, and the air in passing through the nostrils in the act of breathing, brings the odoriferous particles in contact with the olfactory nerves, and the impression is conveyed to the brain and mind, and this causes the sense of smell.

Fred.—O, Father, that explains it very clearly, and I know it is correct, for I recollect I have sometimes been in your office when you were preparing medicines, and I could tell in a moment, by the dust or particles in the air, what kind it was.

Father.—This is generally regarded as the correct theory of the function of smell. It is liable, however, to objections, as it is difficult to explain some facts upon this theory.

Fred.—What objection can there be, Father?

Father.—It is difficult to conceive how particles can escape from metals, and float in the air, and yet metals emit an odor by which different ones can be distinguished. A particle of musk no larger than a pea, will scent a large room for twenty years, and yet not diminish perceptibly in size or weight. A dog will trace the footsteps of his master through the crowded streets of a city where thousands of other feet have trod. The blood-hound will track its game for miles, guided only by the odor it leaves. Now it seems difficult to explain these facts upon the theory I have named. Still, I have no doubt that it is correct, and we are justified in receiving it, although it may imperfectly account for some things, doubtless on account of our limited knowledge.

Use of the sense of smell.

The sense of smell aids us in selecting food, for, as a general rule, food that is healthful has an agreeable odor, while articles which are deleterious to health have an unpleasant one.

In this we see the wisdom and goodness displayed in placing the organ of smell directly over the mouth. For when we have doubts as to the quality of food, we smell it first, and if that does not satisfy our doubts, we then taste it.

Some odors are pleasant, refreshing, and invigorating, and for the time restore the exhausted nervous energy. On the contrary, others are offensive, depress the spirits, and tend to gloom and dejection.

I have spoken of the extreme acuteness of the sense of smell possessed by some of the inferior animals, in reference to the effluvia that emanates from living animals. To these, the sense of smell possesses a far greater degree of importance than to man, for by it they become acquainted with the near approach, both of their prey and their enemy. I have known a dog, stretched upon the floor apparently asleep, to start up and commence barking, as a stranger approached the house, and that too, long before his footsteps could be heard, and ere any of the inmates of the dwelling were conscious of his presence. "It is related of travellers in Africa, that they were always apprised of lions being in their vicinity during the night, by the moans and tremblings of their horses."

To man, the sense of smell is a source of great pleasure, although it might have been far otherwise. The fragrance of ten thousand flowers of the garden and the field, which load the air with their sweetness, and regale, with exquisite pleasure, the sense of smell, might have been to that sense, like the fumes of brimstone or the stench of decomposing animal bodies.

Many other interesting facts might be stated in regard to this sense, but I must now leave the subject for the present, as I have promised to say something upon

THE SENSE OF TASTE.

The TONGUE is the principal organ of taste. There are other parts which participate in the function, though less acutely,—as the lips, the internal surface of the cheeks, the palate, and the upper portions of the œsophagus or gullet. The tongue is composed of muscular fibre, blood-vessels and nerves.

The muscular fibres, which run in almost every possible direction, form the chief part or substance of the tongue. Hence it is capable of great versatility of motions, and can assume, with wonderful rapidity, a great variety of shapes. The tongue is abundantly supplied with blood-vessels, a large artery being distributed upon each side of it. It is also furnished with a great number of nerves, which are filaments from the fifth, ninth, and twelfth pairs. The branch

How taste is produced.

from the fifth pair, called the *Gus'-ta-to-ry Nerve*, is spread upon the upper surface of the tongue and is the nerve of taste. The branches of the ninth and twelfth pairs, are those of motion and sensation.

The tongue is a double organ.

Fred.—Father, what do you mean by that?

Father.—I mean that each side is perfectly distinct in its function, from the other side.

Fred.—How can that be known?

Father.—It is known by the effect of disease upon the tongue. In paralysis, one side may be paralyzed while the other remains perfect.

Frank.—I wish to ask how the sense of taste is produced. Does the substance tasted come in contact with the nerves of taste?

Father.—That is a very proper question, my son, and it is what I intended to explain. The nerves terminate on the surface of the tongue, in what is called papillæ or pimple, which are minute nipples. They are very numerous, and are what produces the rough appearance of the tongue. The end of the nerve is within this nipple, to which it forms a covering to protect it from injury. Liquids, penetrating these papillæ, come in contact with the end of the nerves, and thereby produce the sense of taste.

These papillæ are not equally numerous on all parts of the tongue, being more abundant on the tip, the edges, and the roots, while some portions of the in-

Liquids only can be tasted.

termediate surface are almost destitute of the sensation of taste.

The sense of taste is produced only by liquids. There is no taste in dry food, so long as it remains dry. By mastication, the saliva is mixed with solid food, and in this condition it is brought in contact with the papillæ, and this causes the sense of taste. All substances, therefore, are tasteless which are incapable of being dissolved.

Fred.—How can that be, Father? I have noticed that some metals have a very strong taste. Can they be dissolved?

Father.—Metals are insoluble in water, and the fact that they impart the sense of taste, is explained in this way:—There are certain salts that enter into the composition of the saliva, which are supposed to act upon metals, so as to produce an impression upon the nerves of taste.

When the tongue becomes parched and dry, or when covered with a thick coating, as in fevers, taste becomes imperfect, and in some cases, is wholly suspended. I was once, myself, in this condition, while suffering with a severe fever, when food had no more taste than sawdust.

The sensation of taste is either pleasing or displeasing, and when it is pleasant, we are inclined to swallow, and the sensation is incomplete without it. When the sensation is unpleasant, we at once reject whatever may cause it, and it requires a strong effort of the will to control this impulse, as in the case of taking a nauseous bitter or a pungent stimulant.

Frank.—I know that is true, for I recollect when I was sick I found it almost impossible to swallow some of the medicines you gave me.

Father.—One object of taste is to guide us in the selection of food, and to detect noxious articles which would injure the stomach and endanger life. And had not this sense been perverted and abused, it would doubtless be a correct guide in man, as it seems to be in inferior animals, in which the original design of taste is still answered.

Fred.—How has man perverted the sense of taste?

Father.—By the almost endless admixture of different articles of food, and the use of stimulants and condiments. Also, by alcoholic liquors, and poisonous narcotics, especially tobacco, of which I shall say more at a future conversation. The sense of taste in children is usually acute, and they prefer the mildest kind of food.

The *pleasure* derived from taste may be very agreeable, but it is strictly sensual, and contributes nothing to mental enjoyment or improvement, like that of hearing and seeing.

The number and variety of sensations derived from taste are far greater than those obtained from smell. The epicure is capable of multiplying them almost indefinitely. We can see also, by classifying our sensations of taste, that they are very numerous. We designate many kinds of fruits and medicines by the terms, sweet, bitter, stimulant, astringent, acid, etc., there being many different degrees of taste to each class. The sense of taste was given us not only as a guide in the selection of our food, but if it had not been abused, it would also be a guide to the *quantity* we should take at each meal.

Fred.—How can taste be a guide as to the quantity of food?

Father.—Did you never notice, my son, that the taste is far more acute at the commencement of a meal, when we are in need of food, than it is at the close, when the demand of the system for nourishment has been satisfied?

Fred.—Yes, Father, I am aware of that.

Father.—Well, had our food always been plain and simple, such as nature requires, and had not taste become morbid and unnatural by being stimulated with rich, highly seasoned compounds, taste would be, to a certain extent at least, a guide in the quantity, as well as the quality of food.

God designed the sense of taste to be a source of physical pleasure. But of all the senses, or capacities of enjoyment bestowed upon us, none are so grossly perverted as that of taste.

Frank.—How is it perverted, Father?

Habits formed by using narcotics.

Father.—By stimulating the appetite to excess with rich compounds of food, thereby inducing disease and premature death;—also by creating an unnatural appetite by the use of narcotic poisons, such as tobacco, alcoholic liquors, etc. What can possibly be more offensive and abhorrent to a pure, natural taste than the nauseous tobacco? And yet, strange to say, by use, it not only becomes endurable, but "is rolled as a sweet morsel under the tongue," and about the mouth,—is regarded as exquisitely delicious, and millions become as much wedded to it as though life depended on its use, instead of being destroyed by it.

Did the fact ever occur to you, my children, that articles, the use of which becomes a habit, are always narcotic poisons, and are never healthful beverage or food, demanded by the system for sustenance?

Fred.—I never thought of it; is it so?

Father.—I think it is, my son. Millions, by indulgence, not only create an appetite for tobacco, opium, and intoxicating liquors, but have a constant longing, and craving for them, which is only increased by indulgence. But who ever formed a habit of drinking cold water, so as to have a thirst and craving for it, only when, and in such quantity as the system demands? Or, who ever formed the habit of eating bread "from early morn to dewy eve," and even during a part of the night, and carried a supply in the pockets for the purpose?

Do you not think these facts show the most criminal perversion of this sense, bestowed upon us by our Creator for our safety and enjoyment?

Fred.—I think so, Father.

Father.—The pleasure we derive from the sense of taste is not accidental, neither is it a matter of necessity,—it is wholly the result of God's goodness, and benevolence.

Mary.—Father, how can that be shown?

Father.—The proofs are numerous.

Mary.—Will you please name one?

Father.—It is a necessary condition of life that we eat food,—we cannot live without it. But it is *not* necessary, so far as we know, that the eating of food should be attended with the exquisite pleasure that most persons enjoy. We might have been so constituted that our food would be like sawdust and gravel, or wormwood and gall to the taste, and yet we should be obliged to eat, or die of starvation.

Have you ever thought of this, my children, while eating the delicious peach, pear or strawberry,—and while enjoying the pleasure that such fruits afford, have you raised your thoughts to God in gratitude for these his gifts?

The goodness of God is manifested also in providing the great number and variety of articles suitable for food. The health and growth of our bodies require the use of but a few of them, and during the first year of life we require but one, which is milk, and from this one are formed the bone, muscle, nerve, membrane, and all the parts that compose the body.

Fred.—How many different articles of food are there?

Father.—The number suitable for food is estimated to be more than two thousand. Now if the health and vigor of our bodies require the use of but a few articles, what reason can be assigned why the Creator should provide such a vast number and variety, but his desire to please and gratify the taste he has given us?

Fred.—These things are worthy our serious thoughts, and should certainly lead us to love and adore the Author of all our blessings.

Father.—That is true, my son, and there are other facts in regard to the sense of taste, of which I should like to speak, but have not time at present. I must now say a few words upon

THE SENSE OF TOUCH.

The sense of touch resides in the sensitive nerves. But, as in previous conversations, I have spoken of the principal facts relative to them, I shall now add but little to what I then said. The nerves of touch or feeling, as I have already said, are situated mostly upon the surface of the body, in the true skin. They are placed there, like so many sentinels, to give warning of impending danger, and to convey intelligence to the mind.

The sense of touch—its value.

The chief seat of the sense of touch, as I have before observed, is in the hand, particularly the ends of the fingers, which, for that purpose, are abundantly supplied with the nerves of sensation. All other parts of the body render us sensible or inform us of injury, but they cannot furnish us with definite knowledge of things external. But the hand conveys to us exact knowledge of the qualities of every thing we touch. I mean such qualities as can be known by touch, as heat, cold, rough, smooth, hard, soft, etc. By the hand we also obtain a knowledge of the extension, form, and size of bodies.

There are a variety of pleasurable, and also painful sensations given us by this sense. The sensations produced by electricity and galvanism, hunger and thirst, have been ascribed to it.

This sense is certainly one of the most important given us by our Creator. We can live without seeing or hearing, but how long should we live were we incapable of feeling things that come in contact with our bodies, in the varied occupations of life?

Many have been born destitute of the sense of seeing and hearing. But of the estimate placed by the Creator upon the importance of the sense of touch, we may infer from the fact, that record furnishes no instance in which a human being has been born without it.

This, like all the other senses, was designed by our

Value of the sense of touch.

Creator to be, not only a safeguard to our bodies, and a means of obtaining knowledge, but also a source of pleasure and physical enjoyment.

But had our Creator been a malignant, instead of a benignant being, he might have so formed us that the sense of touch would have been a source of constant annoyance and pain. Everything we touch might have produced a sensation like the irritation of a nettle or the sting of a wasp. The gentle zephyr that now fans the brow and reddens the cheek with the glow of health, might have been like the burning simoom to scorch, suffocate and destroy us.

This sense may be rendered morbidly acute by inflammatory disease, and it may be totally lost by the disease called paralysis. But unless affected by disease, it maintains its integrity, with varying degrees of acuteness, to the very close of life, when this, with all the other faculties, will be lost in death.

I have now, my children, presented to you, the principal facts in regard to the five senses with which God has endowed us. We must now bring to a close this lengthy conversation. But before we part, I wish to recall to your attention one thing to which I briefly referred in a previous conversation when speaking of the nerves of the special sensations,—which is, that the nerves of each of the senses I have described, is specially adapted to its particular function, or office, and *can perform no other*. For example :—Light pro-

Interesting facts.

duces an effect upon the optic nerve, which gives us the sense of vision. But it can have no effect upon the auditory nerve, or any of the other nerves of special sensation. Vibrations, conveyed to the auditory nerve, give the sense of hearing. But vibrations can produce no effect upon the *optic* nerve.

Odorous particles coming in contact with the olfactory nerves, produce the sense of smell. But they can produce no effect upon other organs of sense,—and so of all the special senses. . What is perfectly adapted to produce each particular sense, can have no effect upon either of the other senses.

Frank.—Father, do you mean that the nerves of the different senses are incapable of any other sensation than the special one for which they were designed.

Father.—I do.

Frank.—How then does a strong light thrown suddenly upon the retina of the eye,—as when we look upon the sun,—cause pain in the eye? And how is it that a loud noise, as the discharge of a cannon, causes pain in the ear?

Father.—These pains are produced by the *sensitive* nerves distributed upon the parts, and not by the nerves of special sense.

Frank.—I understand it now, Father, and I can only exclaim,—*it is wonderful!*

Father.—To me, the special senses are among the

most wonderful topics we can contemplate in the structure and endowments of our bodies. And nothing can show *design* more clearly. God saw that each of the senses was necessary, and by a special act of creative power, if I may so speak, formed each of these organs upon a principle, and endowed it with power totally different from either of the others. And the goodness of God in providing us with these senses, can never be fully appreciated until we are deprived of them.

What mines of wealth would not the blind give to regain their sight,—or the deaf to recover their hearing! And yet how little we think of God's goodness in giving us these, while we possess and enjoy them perfectly!

Fred.—It is so, Father. And we feel very thankful for all the intelligence you have given us upon these interesting topics. I trust we shall be more than ever grateful to our heavenly Father for this additional proof of his goodness and benevolent design toward us.

QUESTIONS ON CONVERSATION XIX.

PAGE 280.—What is the organ of the sense of smell?

PAGE 281.—Where is the sense located? Of what is the organ formed? Into what is it divided? What are they called? With what are they lined? What is said of the nerves of smell? What are they called? What is said of the acuteness of the sense of smell? What is said of man? What of beasts of prey? How is the sensation of smell produced?

PAGE 282.—To what is this theory liable? Name the objections. Are we justified in believing it correct?

PAGE 283.—In what does the sense of smell aid us? What do we see in this? What is said of some odors? What of others? Why does the sense of smell possess a greater degree of importance to some animals than to man? Give examples.

PAGE 284.—What is the sense of smell to man? What might it have been? Give an example. What is the principal organ of taste? What is said of other parts? Of what is the tongue composed? What is said of the muscular fibres? Of what is the tongue capable? With what is it supplied? With what is it also furnished?

PAGE 285.—What is the branch from the fifth pair called? Upon what is it spread? What is its function? What is meant by the tongue being a double organ? How can that be known? How is the sense of taste produced? Why are some portions of the tongue nearly destitute of the sensation of taste?

PAGE 286.—In what is there no taste? Why? What is done by mastication? What are tasteless? In what way do metals produce the sense of taste? When does taste become imperfect, or wholly suspended? What more is said of taste?

PAGE 287.—One object in taste is what? How has man perverted the sense of taste? What is said of the sense of taste in children? What of the pleasure derived from taste? What is said of the number and variety of sensations derived from taste?

PAGE 288.—How can we prove that the sensations of taste are numerous? For what was the sense of taste given us? How can taste be a guide as to the proper quantity of food? What did God design the sense of taste to be? What is said of its perversion?

PAGE 289.—How is the sense of taste perverted? What are the articles the use of which becomes a habit? What is farther said upon forming habits?

ANATOMY AND PHYSIOLOGY.—CONVERSATION XIX.

PAGE 290.—What is said of the pleasure we derive from the sense of taste? Name one proof. In what is the goodness of God manifested? Health and growth require what?

PAGE 291.—What is the number of articles suitable for food? Why are so many provided? In what does the sense of touch reside? Where are they mostly situated? For what purpose are they placed there?

PAGE 292.—Where is the chief seat of the sense of touch? What does the hand convey to us? What are given to us by this sense? What have been ascribed to it? What is said of its importance? What proves the estimate placed upon this sense by the Creator?

PAGE 293.—What was this sense designed to be? How might the Creator have formed us? How may this sense be affected by disease? What if it is not affected by disease? What is said of the nerves of special sensation? Give examples. How is pain produced in the eyes and ears?

PAGE 295.—What is said of the value of the senses?

CONVERSATION XX.

MISCELLANEOUS.

PROTECTION GIVEN TO THE BRAIN BY THE FORM OF THE SKULL—EQUALITY IN LENGTH AND SIZE IN DIFFERENT, BUT CORRESPONDING PARTS OF THE BODY—SYMMETRY OF THE BODY PRESERVED BY EQUAL GROWTH OF ALL PARTS—THE FEET, THEIR FORM AND POSITION, THEIR ELASTICITY—THE TOES—THE HAND, ITS STRUCTURE AND ENDOWMENTS—NO TWO PERSONS EXACTLY RESEMBLE EACH OTHER—SAD RESULTS HAD PERFECT SIMILARITY EXISTED—THE RECUPERATIVE PRINCIPLE, OR VITAL FORCE, ITS IMPORTANCE.

Father.—There are many facts of interest, my children, with regard to the structure, formation and endowments of the different parts and organs of the human body, which I have not yet described, as I could not, with propriety, introduce them in the conversations on the special and definite subjects on which we have conversed. I propose at this time, briefly to consider those to which I refer.

Fred.—I feel sure, Father, you will instruct and interest us, whatever subject you may choose. I am surprised at the variety of proof you have already given us of God's goodness in the structure and endowments of our bodies!

Father.—This subject is exhaustless, my son, should we descend to all the minute particulars, which, you recollect, I told you at the commencement, I should not do.

The first I will name at the present time, is the wisdom displayed in the protection given to the delicate and important organ—the brain, by the oval form of the *Cra'-ni-um* (skull).

It is well known that any hollow vessel of an oval or cylindrical form, will sustain without injury, a much heavier blow than a cube or square. A barrel, for example, will sustain a blow that would shiver to atoms a square box of the same material. In like manner the oval form of the skull protects from injury the vital organ it envelops—the brain, far more than it could were it of any other shape.

Also, the bones of the skull composed of two plates, one above the other, with a porous partition between, give very powerful protection to the brain, the outer one being tough and fibrous, and the other dense and very hard. Thus the form and composition of the skull afford the most perfect protection to this vital organ, against any thing but the most unnatural violence.

Fred.—I see it is so, Father.

Father.—Did you ever think, my children, how wonderful it is, that there should be such perfect equality of length and size in the different, but corresponding parts of the body?

To illustrate. Had the bones of the lower limbs been of unequal length, so that one would have been longer than the other, how sad would have been the result,—how dreadful the misfortune! Instead of the present sprightly, graceful movements, we should be obliged to hobble on crutches, or crawl upon the ground. How wonderful, also, that in the growth of the body, the increase of every part should be equal, and thereby perfect symmetry preserved.

Frank.—It is most astonishing!

THE FEET.

Father.—There are also interesting facts with regard to the FEET, which, perhaps has never occurred to you.

Frank.—I presume there are,—will you please describe them.

Father.—I will name some of them, and the first on the form and position of the foot. 1st. The *form* of the foot. The under side, or what is called the sole, is an arch, the extremities of which are the heel and ball. The object or design of this, is clearly to give stability to the standing posture, by the weight of the body resting upon the two extremities of the arch.

Frank.—I see that the form of the feet prevents falling backward or forward, but how does it with respect to a sideway position?

Facts about the feet.

Father.—We are kept from falling sideway's, or to the right or left, by the *position* of the feet,—that is, by the forward part of the feet standing out to the right and left, some fifteen or twenty degrees, in a bracing position, which is doubtless the best and most convenient that could be devised,—and any considerable deviation from it, proves a source of great inconvenience, and is justly regarded as a serious deformity.

Let a person turn his feet in, so that each foot will point directly forward as he walks,—or out, so that they will point to the right and left,—and the instability that will thereby be produced, will enable him to appreciate and value the natural position in which the Creator has placed them.

Another interesting peculiarity of the foot is its elasticity, by which I mean that the large ligament at the heel, called the heel-cord, connected with the powerful muscle in the calf of the leg, is endowed with such force and versatility of motion, that it moves the foot with the celerity of a steel spring, thereby producing that elasticity which contributes greatly to the ease of walking, and the elegance of all the other motions.

The Toes, with their numerous joints and powerful muscular endowments, aid greatly the elasticity of the foot, and render walking a much easier and more elegant performance than it otherwise would be.

Another notable endowment is, that the heel and ball are so constituted and organized that the whole weight of the body, and with the laborer, often one or two hundred pounds additional weight, rests upon the few square inches comprised in them, and yet the feet may be in use every hour of the day without producing any painful sensations.

On what other part of the body so small in extent, could so great a weight rest for several hours at a time, without causing extreme pain and soreness? And was it not goodness that prompted the Creator so to form the feet that they are capable of performing their important office with such agility, and with the absence of all unpleasant sensations?

Fred.—It was. But I confess with shame that I have been using my feet all my lifetime, without ever having thought of what you have referred to.

Father.—Did you ever think, my children, of the atmospheric pressure that rests upon the human body?

Mary.—What is that, Father?

Father.—It is the pressure of the air upon the surface of the body.

Mary.—I did not know that there was any pressure of air upon the body. I never felt any, and I do not see how air can press much upon any thing, it is so thin and light.

Frank.—Nor I either, Father. I never heard of

such a thing before; and certainly I never felt any pressure.

Fred.—Such a thing never occurred to me, nor did I ever hear the subject named before. How much pressure is there, Father?

Father.—The pressure of the air upon every square inch of the earth's surface, and the surface of every thing upon it, is *fifteen pounds*, avoirdupois. Consequently, there is a pressure of fifteen pounds upon every square inch of the surface of our bodies;—and in a man of medium size, the number of square inches of surface is about twenty-five hundred, which, multiplied by fifteen, gives the sum of thirty-seven thousand five hundred. The body of a man, then, bears a constant pressure of *thirty-seven thousand five hundred pounds*.

Fred.—Why, Father! How can that possibly be? Why does it not crush our bodies to atoms? I should not think it possible to live one minute with such a load upon us.

Father.—The reaction of the air within the body, counter-balances the pressure without, and the Creator has so endowed the system, that this pressure, instead of destroying, actually sustains life. Were it removed, we could not live a moment;—were the air taken from the surface of the body instantly, it would be rent into a thousand fragments, by the expansion of the air within the system.

Mary.—Why, Father, you frighten me!

Father.—You need feel no alarm, my child, such a thing will never take place,—and I have not named this fact because there is any practical use to be made of it, but simply to show the wonderful endowments and adaptation of our bodies to the physical conditions of life,—that such an immense pressure can rest upon us every moment, and yet we never feel, nor have any consciousness of it.

Frank.—Are there no means of proving anything upon this subject, excepting what you have named?

Father.—There is one fact worthy of note, which is this:—The air rises forty miles or more, above the earth, and of course, decreases in weight and density, each mile. At an elevation of five miles, the atmospheric pressure is one-eighth less, the effects of which upon the body would be very disastrous, if not fatal. This has occurred in balloon ascensions. Persons at a great elevation, have swooned and become partially or wholly unconscious, with the blood starting from the nose and other parts of the body.

I have previously said something in regard to the structure and endowment of the hand. I wish now to state some additional facts. The hand is a wonderful instrument, or rather set of instruments. An instrument generally is adapted to do but one thing, as a shovel, a saw, a hammer, etc. But how numerous,—how almost infinite are the things the hand is capable of doing!

Let us examine, with more care than we have yet done, the structure of the hand. It has four fingers and one thumb. The fingers are set beside each other on a straight line, and the thumb is on the opposite part of the hand, to match the fingers, thus forming a sort of forceps. This arrangement enables us to grasp and hold with equal ease and firmness, substances varying in size from that of a goose-egg to that of a pebble,—and the sailor a rope of half an inch to two inches in diameter. Now were this changed, and the thumb placed on a line with the fingers, with nothing to match them on the opposite side, how almost worthless the hand would be. Or had the fingers but two joints instead of three, how greatly would their capacity and usefulness be diminished. If you doubt this, to convince yourself you have only to attempt to grasp a substance, bending two joints only of the fingers.

Fred.—I wish, Father, you would tell us more of what the hand is capable of doing.

Father.—It is capable of doing every thing the mind designs; but to enumerate all the hand can do or has done, is impossible. I might sum up in a few words by saying that every thing that has been done in this world, aside from divine agency, has been accomplished by the hands.

What the hands can do.

To enumerate briefly a few things. The hands have felled and cleared the mighty forests, and turned the wilderness into fruitful fields. They have builded railroads and steam engines, cotton mills and steamboats, navies and cities. They have erected the massive structures of granite and marble, whose spires, towering heavenward, direct the mind to the omnipotent One.

The same which can do these mighty acts, can also execute the nicest mechanical and artistic works. The hand that wields the hammer and the sledge, may also be educated to use the writer's pen and the artist's pencil. The Lord's prayer has been written quite legibly, on a surface the size of a dime.

They likewise have made watches, perfect in every part, yet no larger than a diamond that glitters on a lady's ring. It is almost useless to specify. To be able to appreciate, in any tolerable degree, the capacity of the hand, we should visit the World's Fair now open in London, and witness the works there displayed, which, it would seem, almost rival the works of the Infinite.

The hand may also express mental emotions and intentions. Thus, a gentle wave of the hand expresses respect, To point the finger expresses contempt. To clench violently indicates anger. Rapid and violent movements show earnestness and firmness. By beckoning with the hand one can call a person as readily as by the use of the voice. To raise the hands indicates a feeling of wonder and surprise. To spread forth the hands is an expression of devout supplication. To extend the hand on meeting a person, expresses cordial friendship. A parent indicates his desire by pointing to the task he wishes his child to perform,—and in numerous other ways the hand may express what the mind feels and intends.

Frank.—I never realized all this till now.

Father.—Another subject of great interest to which I wish you to give attention, is that no two persons *perfectly resemble* each other.

Mary.—Can that be so?

Father.—Of the one thousand millions who inhabit the earth, no two can be found who are precisely, and in all respects alike. There is often a very strong resemblance it is true, especially in twins, yet a careful scrutiny will show that there are shades of difference, even where the greatest likeness exists.

Mary.—Father, I am surprised at your assertion!

Father.—I think it true, my child. Did you ever see two persons whose appearance was so similar that you could discover no difference, either in the color of the hair, eyes, countenance, or in their form and size, the sound of the voice, or in any other particular?

Mary.—I am not sure that I ever did. But I have

seen twins who were so much alike that I could not readily distinguish them.

Father.—That may be,—the same thing has come under my observation, still, a difference existed which could be discovered by careful scrutiny.

Frank.—Well, Father, what if it be so? What particular importance do you attach to this? I do not see that it is of any great consequence, if there be likeness enough to show that they are all human. Is there any particular design in it?

Father.—I think so, and it seems strange you do not.

Mary.—I am as much at a loss to discover the goodness in this as Frank is.

Father.—I think, my children, you have not thought very deeply upon this subject, for if you had you would have seen, in some degree at least, the kind design of the Creator, which, to me, is very apparent.

Mary.—Please tell us what it is.

Father.—I will endeavor to illustrate. Let us suppose many individuals bear so perfect a resemblance to each other that no difference could be detected. Under these circumstances, you, your mother and myself go into a promiscuous assembly, and I lose sight of you, and in my search I see a girl so very like you that I am sure it is you. I also find a lady who so exactly resembles your mother that I decide she is your mother, and so take you both under my protection. In the mean time the gentleman whose wife I have taken, discovers your mother whom he mistakes for his wife,—the father of the girl also, sees you and believes you to be his daughter.

Mary.—I see that such might be the case.

Father.—If the similarity of appearance I have supposed, existed, communities would be in a state of constant turmoil and strife. Husbands would lose their wives, and wives their husbands, and claim others in their stead,—parents their children and children their parents, brothers their sisters, and sisters their brothers, and no relatives or friends could ever know when they had found their own.

In business relations, the creditor could not identify the debtor, nor the debtor the creditor. In legal processes, evidence in courts of justice would be worthless, and if allowed in criminal cases, might condemn the innocent to death, and clear the murderer from deserved punishment. In short, such would unsettle the foundations of society, and fill the world with anarchy and confusion.

Frank.—I now see there is a wise and good design in the dissimilarity observable in the human species, for which deep gratitude is due the Creator.

Father.—From the facts cited, we discover the extreme *perfection* of God's works, as well as his wisdom and goodness. That he has formed so many millions of beings, bearing so strong resemblance that they are never mistaken for another species, and yet **no** two can be found who are perfectly alike, is most astonishing, the consideration of which may well fill the mind with thankfulness and reverence.

Mary.—I think so too, Father.

Father.—During these conversations I have spoken on the endowments of many organs and members of the body. I now wish to call attention to a capacity or power bestowed upon us by our Creator, totally different from any thing I have yet named,—which is not limited to any one, or any number of organs, but pervades alike every part and fibre of the body. I refer to the *recuperative force,* or the *self-healing, vital principle.*

Fred.—What is that, Father?

Father.—It is the principle by which injuries of the body are repaired, and diseases cured,—such as cuts, bruises, burns, broken bones, dislocated joints, fevers, rheumatisms, etc.

Fred.—I thought the cure of these was the business of the surgeon and physician.

Father.—The surgeon may dress a wound, set a broken bone, and reduce a laxation,—and the physician may administer appropriate medicines for disease, but all this is not curing the patient, it is only *aiding* this recuperative principle.

Fred.—What is this principle?

The vital principle.

Father.—I do not know, my son. What it is, has never been defined, for, like its Author, it is incomprehensible and undefinable. *What it does* is all we can know of it. In case of a flesh wound, as a cut or a bruise, this vital principle, by its mysterious operations, fills up the wound with new flesh, and restores the part to soundness. In case of a broken bone, it throws out a substance which unites and firmly cements the two parts, and restores the bone to its original strength,—but *how* it effects these objects no one knows.

In severe diseases, the physician sometimes administers every medicine his skill and judgment dictate, but the disease is not subdued, and the patient is finally given up to die.

In this condition, when the scales of life and death seemed to hang at equipoise, and vibrate with awful uncertainty, then this recuperative force—this vital power, seized the fell monster, disease, thrust him from the citadel of life, and health preponderates in the scale.

Frank.—Can the physician do nothing in the cure of disease, independent of this vital force?

Father.—Nothing.

Frank.—Is it this power alone that cures disease and heals wounds?

Father.—It is, my son.

Frank.—What benefit then are physicians?

Vital force and vital action.

Father.—To answer your question fully, would be to explain the principles of medical science, which I have not proposed to do. I will, however, state a few things which may interest you. Every fibre of the body, during each moment of life, is pervaded by, and is under the influence of this vital force, producing what is called *vital action.*

Frank.—What is vital action?

Father.—It is the operations which sustain the body in a healthy condition, as the action of the nutritive vessels, the absorbents, the secretory organs, the excretory ducts, the capillaries, etc.

I do not mean to say that these are the vital force,—they are simply the effect or result of it. The cessation or derangement of any of these vital processes, soon prostrates the health, and, if long continued, destroys life.

The integrity of these vital processes which sustain life and health, depends upon the undimished force of the vital principle. This force, however, varies greatly in different persons, and in the same under different circumstances. When the vital force and action, from whatever cause, are diminished, it causes, or predisposes to disease. But this vital force is capable, when diminished or partially exhausted, of being excited to increased action by numerous agents called medicine.

Now the skill of the physician, and his ability to

Medicine can act only on living bodies.

benefit the sick, depend upon his knowledge and use of such agents and processes as will control this vital action,—stimulate it when too low or too weak, and modify and control it when unequal and deranged.

Mary.—I am surprised, Father, at your statement as to the effect of medicine. I supposed it had the power of itself, to act upon the system, independent of every thing else.

Father.—That is a very common, but very great error, my daughter. Were not medicines dependent for their action upon this vital force, they would as readily act upon a dead as a living person.

We must now close our present interview, and with it I shall conclude my description of the structure and functions of the human body.

I had intended to speak of other parts, as the face, the organs and faculty of speech, etc., but I must defer a further description to some future period, as I have already transcended my original intention.

Children.—We regret, Father, to hear you say that this is to be our last conversation, for we have been highly delighted and greatly instructed, and we feel truly grateful to you.

Father.—I do not say, that this is to be our last conversation, for I have more to impart, but what I shall further say will not be a description of the organs and functions of the body, although I trust it will not be less interesting and instructive than what has been said.

QUESTIONS ON CONVERSATION XX.

PAGE 297.—What gives protection to the brain? Explain how the oval form of the skull protects the brain? What is said of the bones of the skull? What is said of the equality in length and size of different parts of the body?

PAGE 298.—Give an illustration. What produces symmetry of form? What is said of the form of the feet? What is evidently the design of this?

PAGE 299.—How are we kept from falling sideways? What will produce instability? What is said of the elasticity of the foot? What is said of the toes?

PAGE 300.—What other endowment of the foot is named? Is the goodness of the Creator manifested in this?

PAGE 301.—What is the pressure of the air upon every square inch of the earth's surface and everything upon it? The body of a man then bears how much pressure? Why does it not crush him to atoms? Were this pressure removed what would be the effect?

PAGE 302.—What fact is worthy of note? What is said of the hand?

PAGE 303.—What of its structure? What does this arrangement enable us to do? What is the hand capable of doing? What have the hands done?

PAGE 304.—What else can they execute? Can mental emotions be expressed by the hand? What expresses respect? What contempt? What anger? What firmness? What can be done by beckoning?

PAGE 305.—What indicates surprise? What supplication? What expresses friendship? How does a parent indicate his wish? What is said of personal resemblance?

PAGE 306.—Is there a kind design of the Creator in this? Give illustrations.

PAGE 307.—What would be the result upon communities of a perfect similarity? What upon business relations? What in legal processes? What upon society? What do we discover from these facts?

PAGE 308.—What force or principle is here spoken of? What is done by this principle? What is said of the surgeon and physician?

PAGE 309.—Do we know what this principle is? What is known of it? What does it do?

PAGE 310.—By what is every fibre of the body pervaded? What is vital action? What more is said of the vital principle and vital processes?

PAGE 311.—On what does the skill of the physician depend? What is said of the action of medicine?

CONVERSATION XXI.

THE SOUL.*

MAN A COMPOUND BEING, CONSISTING OF BODY AND SOUL—THE SOUL OR SPIRIT INFINITELY SUPERIOR TO THE BODY—ALL POWER RESIDES IN SPIRIT—PROOFS OF THE EXISTENCE OF THE SOUL OR SPIRIT—OUR KNOWLEDGE OF SPIRIT AS POSITIVE AS THAT OF MATTER—THE CAPACITIES OF THE SOUL—ITS INFINITE VALUE.

Father.—The previous conversations have been devoted to a consideration of man's physical system. I have briefly described the structure, functions, and endowments of the principal parts and organs of the body, and have enumerated some of the many proofs of God's benevolent design in its structure.

I now wish to direct your attention to the weighty fact, that man is a compound being, consisting of the body which I have described, and a soul or spirit, which is an immaterial and immortal principle, and is the seat of the different affections and passions, such as love, hatred, anger, etc.

We have seen that the body is a most wonderful

---

* The author does not claim entire originality of thought for some portions of the present conversation. For the sentiment that the manifestation of the properties of spirit, is proof of its existence, credit is due the author of "A Pastor's Sketches."

piece of mechanism, every part of which proclaims its divine origin. But with all its perfections it is the lowest and most inferior part of man. The spirit or soul as far surpasses the body, as the splendors of the noon-day sun exceed midnight darkness, or as the infinite excels the finite.

Frank.—Why is the soul superior to the body?

Father.—Because the body is composed of matter, one of the properties of which is inactivity or inertia.

Mary.—What is inertia?

Father.—It is the inability of matter when at rest to put itself in motion, or when in motion to assume a state of rest. Our bodies, therefore, are as incapable of motion as a stone or a clod of earth.

Mary.—It seems to me, Father, that your statements do not agree. Did you not tell us when speaking of the muscles, that they possess great power, and produce all the movements of the body?

Father.—I did, but I told you also that the voluntary muscles had no power of motion independent of the mind, and that all their movements were under the control of the will?

Fred.—If the muscles do not possess the power that moves the body, where does it reside?

Father.—In spirit or mind. The power that moves a muscle does not reside in the muscle itself. The muscle is only an instrument, which must be put in motion by a power outside of, and distinct from itself, and that power resides in spirit not in matter. Our bodies are totally incapable of motion, and are utterly destitute of power, independent of the spirit. Indeed, all power in the universe resides in spirit or mind.

Fred.—Why, Father! Is it really true, that there is no power except in mind or spirit? I recollect that much is said in my Natural Philosophy, in regard to the mechanical powers,—the lever, the screw, the wedge, the pulley and some others. Now what do you say to these?

Father.—I say they possess no power whatever, independent of mind. Did you ever see them do any thing alone? They can do nothing because they have no mind or spirit. It was the mind that contrived these mechanical powers; but it could not invest them with the ability to work alone. The mind employs them to do what it could not accomplish without. Not only all power, but as a result, all motion resides in spirit. With my mind, or spirit, I move, as I please, the matter of which my body is composed. God, the Infinite Spirit, moves as he pleases, the universe of matter. But *how* my spirit moves my body,—or how the Infinite Spirit moves the millions of worlds that compose the universe, is a profound and inscrutable mystery.

Fred.—Can this be so, Father?

Father.—It is so, my son. This body, with all its endowments, is only a wonderful instrument made for the mind to work with. It is the mind that builds houses, and ships,—that cultivates the soil, and navigates the ocean,—using the body as an instrument, with which to accomplish these objects. Again, the body is dependent upon the mind for every thing, the mind acting the part of a guardian,—protecting it from harm and danger,—providing it with food in health, and medicine in sickness,—with house to shield it from the inclemency of the weather, and with clothing and fire to keep it warm, thus prolonging the life, which, without this care, would soon come to a close.

Frank.—Then all the power the body possesses belongs to the spirit, and all the labor it does is accomplished by it. I wish you could tell us, Father, how the mind can exercise such control over the body?

Father.—This, as I have before said, is one of the profound mysteries in the works of Deity. That it is so we know. My mind moves my body,—its mandate is instantly obeyed. But *how* mind acts upon matter, none but the omniscient One can tell. It may, perhaps, be feebly illustrated by the locomotive steam engine. When we examine its massive iron shafts and wheels, and its bars of steel,—and see it in full motion, moving with great speed, a train of cars laden with hundreds of tons burden, we are amazed and exclaim,—what a powerful engine! But does the power that produces such results, reside in the engine?

Take away the steam, which, like spirit is invisible, and how soon its motions cease, and how powerless it is. And so it is with man's physical frame, which, as Job says, is "fenced with bone and sinews," and is furnished with powerful muscles. When God takes away the spirit, all power immediately ceases.

Mary.—How can you say, Father, that steam is invisible, when we often see volumes of it rising from steam engines?

Father.—My daughter, that to which you refer, is not steam, it is vapor, and is produced by steam coming in contact with atmospheric air, which condenses it to vapor. Steam is no more visible than the air we breathe.

Mary.—I understand it now, but I did not know before that steam was invisible.

Fred.—I do not question the truth of what you have said, Father, but it seems strange to me, if such power resides in spirit, that we should know so little about it. Every body has knowledge of matter, but how little is known of spirit.

Frank.—That is what I have been thinking, Father, as I have listened to what you have said to Frederick. If spirit possesses such power, and is so superior to the body, why is it that so little is known about it? It is strange that what is so important should be so obscure.

Mary.—I am of the same opinion. I should suppose that God's goodness would have prompted him to give us the most knowledge on what is most important. But the reverse seems to be the case, for we know much more about matter, and have more certain knowledge of it, than of spirit.

Father.—What do you mean, my child, by having more certain knowledge of matter than of mind or spirit?

Mary.—I mean we know for *certainty* that matter exists,—for we see it, and feel it,—we recognize it with all the senses, and therefore we *know* that it does exist; but we have no such proof of the existence of spirit.

Father.—My children, let us examine this subject with great care and candor, for it is one of great moment.

You say we have no such proof of the existence of spirit or mind as we have of matter. I admit that we have not the same kind of evidence, but I think it not less positive. Indeed, the fact that we know any thing about matter, is proof of the existence of mind, for mind only has knowledge,—we ascertain truth with the spirit, not with the body; and if we acquire any knowledge about the body, we are sure we must have a mind which possesses that knowledge. We can make no progress in the knowledge of matter, without at the same time advancing in the knowledge of mind.

The five senses prove the existence of spirit.

You say also, we are acquainted with matter by the special senses,—seeing, hearing, feeling, tasting and smelling, but the senses give us no knowledge of the existence of spirit. Now I think the *reverse* is true. I believe the five senses prove the existence of spirit more fully than that of matter.

Fred.—How so, Father?

Father.—The senses cannot *exist* independent of spirit. We see, hear, feel, taste and smell with the mind. Therefore the senses prove mind more than matter.

Fred.—I should be pleased to know what proof can be given that we see, hear, feel, taste and smell with the mind.

Father.—If the communication between the brain, —the seat of the mind, and each organ of sense be cut off by ligating, (tying) the nerve, the sense is destroyed. No sensation of any kind can possibly be produced. As a result, we should be in total darkness—should be perfectly deaf—could feel neither fire nor frost—could not smell the sweetest perfumes, nor taste the richest viands. Now is not this positive proof of the existence of mind or soul?

Fred.—It certainly is, but I think I should never have discovered it.

Father.—Another proof is, its action upon the body. A *fright* causes the blood to recede from the surface, and the countenance to turn pale. It also effects the muscles, producing weakness and trembling of the whole system.

The *state* or *emotions* of the mind in which we habitually indulge, fixes its impress clearly upon the countenance. Thus love gives a cheerful, happy expression, which sparkles in the eyes, and plays in smiles upon the lips,—anger, a morose, surly, sour, sulky look, which darkens the brow and renders the expression of the face repulsive,—and so of all other emotions, as sympathy, sorrow, anxiety, pity, etc., each leaving its own peculiar impression, so clearly depicted upon the countenance, that it affords a sure index to the state of the mind which produces it.

The *business man*, who suddenly perceives his affairs are so embarrassed as to threaten bankruptcy, may thereby be made to tremble like a leaf. But his mind alone effects his body. He has not been seized by the sheriff, nor locked up in jail. No one, perhaps, but himself, has any knowledge of his affairs. Yet the certainty that he is a bankrupt, causes him to tremble and turn pale. A *criminal*, under a sense of his guilt, which is known only to himself, shakes like a reed in the wind. Yet nothing has touched his body,—the law has not yet laid its strong hand upon him,—he is not thrust into a felon's cell; yet the spirit shakes the body with wild convulsions. Now does not this demonstrate the truth of the inspired declaration, " There is a spirit in man?"

In the last examples, the effect upon the body is purely mental. Neither of the bodily senses are affected; it is reflection and consciousness alone that thus powerfully affects the material frame. Now when a mere thought deranges the circulation of the blood, shatters the nerves, and shakes the whole material system, and that too, without touching one of the bodily senses,—what must we think of such a power? Do we not know that it cannot be the effect of matter, and must be the action of spirit?

Fred.—I think so, Father; and yet I cannot rid myself of the impression that we have more certain knowledge of matter than of spirit.

Father.—What do we know of matter that we do not of spirit?

Fred.—We cannot tell what spirit is.

Frank.—And is it not the same with respect to matter? We know as much of the one as the other. You may again reply that we can see and feel matter. But I have proved that it is the mind or spirit that sees and feels, and therefore, what you regard as positive proof of the existence of matter, is equally positive with respect to the existence of spirit.

Fred.—But did you not say when speaking of the nerves, that all the knowledge we have of the universe is obtained through the senses?

Father.—I did, and I repeat the same now. But as I have before stated, if we were destitute of mind

Matter and spirit known only by their properties.

or spirit, we should be of the senses also. And as to the *certainty* of what is obtained by the senses, I wish to ask if you are more certain of what you see, than you are of the feeling of love, hatred, anger, etc. Is not the knowledge of the one as positive as that of the other, though obtained in a different way? As to the *extent* of our knowledge it is *greater* about spirit than matter.

Fred.—Will you please give us your reasons for such a belief?

Father.—All we can know, either of matter or spirit, is their properties or qualities,—we know nothing of their essence. Now let us enumerate the known properties of each. Those of matter are weight, figure or form, color, hardness, impenetrability, divisibility, porosity, and inertia or inactivity. We know that spirit has the following properties or capacities, to wit:—It loves, hates, reasons, remembers, perceives, judges, compares, wills, fancies, has imagination, consciousness, and is capable of suffering pain or enjoying pleasure. Thus you see we know much more about spirit than matter, as its properties are more numerous.

Fred.—I shall never again indulge any doubts on this subject.

Father.—The Mosaic history of the creation of man and animals, is full of interesting instruction, and throws much light upon this subject. According to that history, God first created the inhabitants of the waters, then " cattle and creeping things and beasts of the earth." But all these were irrational creatures,—they possessed no soul. They had not the power of thought, reason and control. A superior being was still needed,—as is beautifully expressed by Dryden,—

> " A creature of a more exalted kind,
> Was wanting yet, and then was MAN designed,
> Conscious of thought, of more capacious breast,
> For empire formed, and fit to rule the rest.
> Whether with particles of heavenly fire
> The God of nature did his soul inspire,
> Or earth but new divided from the sky,
> Which still retained th' ethereal energy.—
> Thus while the mute creation *downward* bend
> *Their sight*, and to their earthly mother tend,
> Man *looks aloft*, and with *erected eyes*,
> Beholds his own hereditary skies."

The sacred historian, after a description of the formation of the body of Adam, says, " and God breathed into his nostrils the breath of life, and man became a living soul." Or as the original is,—" the breath of lives," that is, animal and spiritual life.

The body was formed, perfect in every part, yet it had no life, for the reason that it had no spirit or soul. Not only Moses, but the Scriptures from beginning to the end, speaks of man as a compound being,—the soul it calls the inner or hidden, and the body, the outer or visible man.

The body grows and arrives at a state of maturity and perfection; its powers fully develop; it then decays and dies. But the mind or soul may advance in

Capacities of mind unlimited. Endless progress of the soul.

knowledge and virtue, until it vies with angels. Indeed, the capacities of the mind seem to be unlimited. The more knowledge it obtains, the better qualified it is for farther attainments. Persons in the pursuit of science never feel that their present acquirements, however great, disqualify them for still greater advancement. And during the endless ages of eternity, the soul will be progressing in knowledge of the works of God,—nor will the time ever come when the limit of its capacity for improvement shall have been reached, but it will ever be approximating its Maker by greater, and still greater degrees of resemblance. Yet the soul will always be the *finite*, and God the *infinite*.

Fred.—How can that be, Father? If the soul will be forever progressing in knowledge, and in resemblance to the Creator, will not the time arrive in the remote ages of eternity, when the soul will, at least, have made a near approach to its Maker?

Father.—Endless progress of the soul in knowledge and goodness, will, of course, increase its resemblance to the Creator. But the soul, in reference to God, will always be like one of those geometrical lines which may approach another forever without the possibility of touching. The soul is not only capable of endless progress in knowledge, which is a purely mental quality,—but its moral qualities are not less wonderful. Although naturally impure, depraved, and vile, yet by the exercise of its own free agency, and the aid of the Holy Spirit, it may rise in the scale of moral excellence, to the perfection and purity of angelic natures. On the contrary, the soul, by yielding to its naturally depraved and corrupt propensities, is capable of sinking to a state of degradation and misery which is frightful to contemplate.

Frank.—You surprise me, Father! Our physical structure—the body, about which you have told us so many wonderful things, it seems, is of very little consequence compared with the soul.

Father.—It is, my son. And you may be assured that the value of the soul can never be over-estimated.

In previous conversations we have seen the exquisite workmanship, and extreme perfection evinced in the structure of these bodies which are soon to die and return to their original elements. Now if the structure which is to continue but a few years, evinces such skill, what must we think of the perfection displayed in the creation of the soul, which is not only endowed with powers approximating infinity, but whose progressive existence is to be coequal with Jehovah!

This endless durability of the soul may give us some idea of its value. In mechanics, we estimate the worth of things, in a great measure, by their durability. Thus, if a steam engine, which is usually worn out with eight or ten years service, had the capacity of enduring the wear of a thousand years, how greatly would its value be enhanced. So if the soul, with its untold, inconceivable, and ever increasing capacities, is destined to an endless existence; what archangel tongue can proclaim its value!

Fred.—What proof have we, Father, that the soul will exist forever, or even that it will survive the dissolution of the body?

Father.—The proofs are various. The fact that the soul is immaterial, and consequently indestructible, is regarded by Theologians as proof of its immortality. But the book of divine revelation—the Bible, contains the most direct and satisfactory proof. Numerous passages might be cited, but I shall quote but two, and these are of undoubted authority. Solomon, speaking of the death of the body, says, "Then shall the dust return to the earth as it was, and the spirit shall return unto God who gave it." Christ says,—"Fear not them that kill the body but cannot kill the soul." What can be more direct proof that the soul lives after the body is dead?

We must now close this conversation, and in doing so I would ask, what can be of more thrilling interest to man than the theme upon which we have conversed. If the life we live in the body is but an embryo existence—an initiative state, which, at its dissolution, shall burst into the higher life of celestial beings,—the soul sustaining the moral qualities and

Reflections.

character obtained in this life ;—what solemn importance attaches to the present state of existence!

May we arise from the consideration of this subject, with a high and fixed purpose that our lives in future shall be exalted to a conformity with the divine purpose in giving us this earthly existence, and, with its due improvement, a prospective existence of unending felicity.

QUESTIONS ON CONVERSATION XXI.

PAGE 312.—What weighty fact is here stated?

PAGE 313.—What is said of the body? What of the spirit or soul? Why is the spirit superior to the body? What is inertia? Where does the power of motion reside? What is said of the muscle?

PAGE 314.—What is said of the mechanical powers? What does my spirit do? What does the Infinite Spirit do? What is said of this body?

PAGE 315.—What is said of the mind? Can you tell how mind acts upon matter? What illustration is given?

PAGE 316.—What is said of steam? What more is said about matter and spirit?

PAGE 317.—What is proof of the existence of mind or spirit? How do we ascertain truth or facts?

PAGE 318.—What do the five senses prove? Do we see, hear, feel, taste, and smell with the mind? What is the proof? What is another proof of the existence of spirit?

PAGE 319.—What do the emotions of the mind do? Give examples. What is said of the business man? What of the criminal? What does this demonstrate?

PAGE 320.—What is said of the last examples? What must we think of such a power? Do we know as much of spirit as of matter? Can we tell what either is? What if we were destitute of mind or spirit?

PAGE 321.—What is said as to the *certainty* of knowledge obtained by the senses? What as to the *extent* of our knowledge of spirit and matter? What is all that can be known of either spirit or matter? Can we know anything of their essence? What are the properties of matter? What are the properties or capacities of spirit? The properties of which are more numerous? What is said of the Mosaic history of the creation of man and animals?

PAGE 322.—What was created first? What next? What is said of them? What was still needed? What is said in regard to Adam? Why had his body no life? What do the Scriptures say of man? What is said of the growth, maturity, and decay of the body? What is said of the mind, or soul?

PAGE 323.—What is said of its capacities? What of persons in pursuit of science? What of the soul's endless progress in knowledge? What will the soul, in reference to God, always be like? What is said of the moral qualities of the soul?

PAGE 324.—To what may it rise in moral excellence? To what state is it capable of sinking? What is said of the endowments and progressive existence of the soul? What gives us some idea of its value? By what do we estimate the worth of things in mechanics? Give an example.

PAGE 325.—What, then, is the value of the soul, if estimated by its endless existence? What proofs are there that the soul will exist forever, or even survive the death of the body? What gives solemn importance to the present state of existence?

CONVERSATION XXII.

LAW.

ALL THINGS UNDER THE CONTROL OF LAW—LAW DEFINED—HEALTH SECURED BY OBEDIENCE TO THE LAWS OF LIFE—DISEASE THE RESULT OF VIOLATED LAW—HEALTH AND DISEASE DEFINED—THE PHYSICAL, MENTAL AND MORAL EFFECTS OF DISEASE—ALL SUFFERING THE RESULT OF VIOLATED LAW.

Father.—My dear children, I have given you a brief description of man's physical structure, and have presented some of the proofs that he possesses an immaterial and immortal part, called the spirit or soul, between which and the body there exists a most intimate, though mysterious union, by which they are capable of acting and reacting most powerfully upon each other. And although I think I have redeemed the promise I gave you at the commencement, yet I feel that my instructions would be very imperfect should I stop here.

Fred.—We shall be glad, Father, to hear anything further you may have to say, for I am sure we shall be both interested and instructed.

Father.—A knowledge of the structure of the body,

and the functions of its numerous organs, is of thrilling interest to all thoughtful persons. But to understand the best means of preserving the health, is by far the most important knowledge we can possess, in relation to it. The former is purely scientific, making us acquainted with the wisdom, perfection and benevolence of God,—the latter is not only scientific, but is of the highest practical importance. Have you duly considered, my children, the vast value of bodily health, and the means necessary to its preservation?

Frank.—I think health is very desirable, for one feels very unhappy when suffering pain, and when confined with sickness. But what are *we* to do? If God sends sickness and pain, I should like to know how we can prevent it. Can we thwart his plans, and counteract his designs? Disease always has existed, and doubtless always will.

Fred.—Is it not evident, Father, that God designed there should be pain and disease,—for people must die some time, and what should cause death were there no sickness?

Father.—You think, then, that sickness, pain and death take place because God wills it should be so; that when a man dies, it is because God says, in reference to him, "That man has lived long enough, and now I will kill him," and so sends disease, which, in a few weeks or months terminates his life, or else He cuts him down at a blow with paralysis or apoplexy.

False sentiments about health.

Fred.—I suppose it must be so,—we often hear it said it is a great mystery why God should take away those whose lives are so important, as parents from dependent children, ministers from their people, children from aged parents upon whom they depend for support. And I recollect, too, that the minister said, at the funeral I attended a few days since,—that it was a mystery which we could not now comprehend, why a man so useful in the community, and so much needed in his family, should be so suddenly taken away,—" But," he added, " God has done it, and we must be reconciled."

Father.—Were we to credit what is often asserted, we should believe that health and sickness are mere things of accident,—that one possesses good health, and another is feeble and diseased because it happens to be so, or the one chances to have a better constitution than the other; or else it is so for the reason that God, in his infinite wisdom, saw it was best it should be so, and in the operation of his mysterious providence, so decreed it. But is not such a sentiment " charging God foolishly?"

Can we suppose God has created this wonderful structure, in which we trace such marked proofs of wisdom and goodness, to be the mere sport of chance, —sometimes sick, at other times in health,—some persons with firm, others with feeble constitutions, without any appreciable cause? We will not degrade ourselves, nor dishonor our adorable Creator with such a sentiment!

Fred.—We should be glad, Father, to know the truth upon this subject. Will you please instruct us.

Father.—I will state a few facts. Every thing God has made, animate and inanimate, matter and mind, is under the control of law.

Mary.—What do you mean by law? I thought law is what we find printed in books, made by legislators to control the action of communities.

Father.—Law is a term of broad significance, and is applicable to a great many subjects, and takes its name in accordance with the nature of the subject to which it is applied.

Statute law, is the term applied to what you have described, and is, as you say, designed to control communities, and has, like all law prescribed to rational creatures, penalties annexed to its violations.

Moral law, is that which prescribes to men their religious and social duties, that is, their duties to God and to each other.

Physical law, or *law* of *Nature*, is that which controls all matter and motion, whether it be the rushing cataract, or the fall of the gentle shower, or the more gentle dew,—the fearful tornado or the gentle zephyr, —the sweeping of worlds through infinite space, or the floating of sunbeams, rendered visible only by the sun's rays. Even the forked lightning, which seems to dart at random athwart the skies, is under the control of law.

Laws of vegetation, "are the principles by which plants are produced, and their growth carried on till they arrive to perfection."

Then there are Municipal laws, laws of Nations, the Mosaic law, laws of Gravitation and Cohesion, etc. But those to which I wish to call your special attention, and which particularly belong to our subject, are called *the laws of animal nature, or the laws of life.*

Mary.—What are they, Father?

Father.—They are the principles by which the functions of animal bodies are performed, such as respiration, the circulation of the blood, digestion, nutrition, the various secretions,—in short, they are those which control the action or function of every part, organ and fibre, of the entire body.

It is a principle, as already stated, established by the Creator for the protection of life and health, the knowledge and observance of which secures health, and the violation produces disease. What the observance, and what the violation of the laws of life and health are, I shall explain in future conversations.

Here, my children, is a solution of the problem why one person is sickly and another healthy,—why one droops and dies in early life, and another lives in the possession of firm health to a good old age. It is chiefly because one obeys the laws of health, and the other violates them. One allows the wheels of life to run on smoothly, without friction or wearing, the other applies corrosives, by which the machinery is suddenly stopped, or is soon worn out.

Fred.—Do you mean to say, Father, that we should always enjoy health, and should live to old age, if we carefully observed, and never violated the laws of life and health?

Father.—I have no doubt that such would be the case, *casualities* excepted, provided we were born free from the taint of disease, or hereditary predisposition to it. Old age is the only natural death of man;—when life closes by self limitation, or by the exhaustion of the natural and vital forces.

Frank.—Why, Father, you surprise me! Is the death of so many millions of infants and children, of youths and of the middle aged, the result of the violation of the laws of life?

Father.—I believe it is, with the exceptions I have made.

Fred.—This is a startling announcement, Father!

Father.—I think it true, nevertheless.

Fred.—And you think every one would live to old age, were they to obey the laws of health?

Father.—I do, if, as I have already said, all were born free from the taint of disease. But in man's present deranged and depraved physical condition, I

Disease and pain the result of violated law.

believe it impossible for all to obey the laws of health. Indeed, no small portion of mankind came into the world with the seeds of disease deeply sown in their constitution,—such as scrofula, consumption, etc., which soon germinate and produce premature death. But in such cases, disease and death are the result of violated law, as much as in any other case; not of the sufferer, it is true, but of his ancestors; consequently, no obedience to law that he can render, can avert the penalty, for it cannot reach back to the original cause. But deducting all such cases, there still remain vast numbers who die annually, whose death is as absolutely the result of violated law, as if it were produced by severing the jugular vein, or thrusting a sword through the heart. For, let it ever be remembered, it is an utter impossibility to violate any of the laws that God has established, without sooner or later suffering the penalty.

The early history of the human race, furnishes a fact which I think goes far to sustain the views I have expressed,—a fact, which, perhaps you have never seriously considered.

Fred.—What is it, Father?

Father.—It is the following:—According to the Mosaic history of generations which lived before the flood, (See Genesis, chapter 5th), children did not die before their parents, as is so common in our time. It is expressly stated that fathers died and left their children, who arriving at manhood, became fathers, and in their turn passed from the stage of life, and left their children to follow them in like manner. Thus successive generations came upon the earth and passed away, parents leaving their children,—but in no instance do we read of children dying before their parents; whereas now, it is no uncommon thing for whole families of children to be swept away by death, and the parents left to mourn their sad bereavement. Then again, consider to what a vast age people lived in those early times;—usually several hundred years.

Frank.—How do you account for these facts, Father?

Father.—I know of but one way to account for them, which is, that man came from the hands of his Maker with a pure and perfect body, free from all taint of disease, and endowed with a degree of vital force which carried the wheels of life smoothly on for nearly a thousand years. I know of no reason to believe that scrofula, fever, rheumatism, consumption, gout, neuralgia, dyspepsia, fits, cancer, apoplexy, palsy, cholera, and the numerous less prominent diseases were known in those early times. These, and man's present degenerate physical condition, are the result of violated physical law.

Mary.—I hope, Father, you will teach us how to observe the laws of life, and give us instruction as to the best means of preserving our health.

Health and disease defined.

Father.—I intend to do so. But I wish first to state a few facts to impress upon your minds the *inestimable value* of health. In doing so I shall endeavor to point out some of the effects of disease, upon our happiness, usefulness, and well-being, in a *physical, mental* and *moral* or *religious* point of view. But that we may talk intelligibly upon this important subject, let us first define the terms *health* and *disease*.

Health is that state of a living animal body, in which the parts are well organized and sound, and in which they all perform freely their natural functions. In this state a person or animal feels no pain.

Disease, in its primary sense, is pain, distress, uneasiness. But these are rather the effects of disease, than disease itself.

Disease properly, means any deviation from health, in function or structure; the cause of pain or distress; it is that state of a living body in which the natural functions of the organs are interrupted or disturbed. A disease may effect the whole, or a particular part of the body, as a limb, the head, heart, stomach, lungs, liver, bowels, etc., and such are called local diseases. The first effect of disease, or this derangement of the vital functions, is uneasiness and pain, and the ultimate effect is total cessation of the functions, which is death.

The *physical* effects of disease are numerous, some of the most prominent of which I will name. As I

Physical effects of disease.

have said, pain is the first effect of disease; and it is usually that which gives the first intimation of its existence. And this is produced in various degrees of intensity, according to the nature of the disease. Some diseases, as rheumatism, neuralgia, colic, gout, etc., are attended with acute pain, usually of extreme severity, while many forms produce a dull, heavy ache, accompanied with nervous excitement, debility and languor. Persons are thus often deprived of all rest, by day and by night, for weeks and months and sometimes for years. Such can say with Job, "Wearisome nights are appointed to me. When I lie down, I say, when shall I arise and the night be gone? and I am full of tossings to and fro unto the dawning of the day."

Some diseases seem to be beyond the reach of remedial means, preying upon and wearing away its victim by slow degrees, until death terminates his sufferings. Again, some forms of disease produce incurable deformity. *Rheumatism* and *hip-disease*, often draw the joints out of place, and so alter their form and size, that no surgeon can replace them. *Spinal disease*, produces contortions of the spinal column, which usually remain for life. *Palsy* causes a cessation of sensation and of motion. In *Locked-jaw*, the muscles of the lower jaw become so contracted and hard, that the mouth cannot be opened, and in some cases it is difficult, if not impossible to pry it open. Also, the muscles of the neck and back, and indeed, sometimes of the whole body, are affected with violent spasms.

Persons suffering with the diseases I have named, may in truth exclaim with the poet,—

> "O life, thou art a galling load,
> A long, a rough, a weary road,
> To wretches such as I."

Disease not only causes pain and deformity, thereby disqualifying its victim for all the comforts of life, but it incapacitates for the *duties* of life. How can the farmer, in any one of the conditions I have cited, attend to his agricultural pursuits? or the mechanic to his mechanical operations, or the merchant to his mercantile affairs, or how can the mariner navigate the ocean, and steer his bark over its rough and surging billows? So far from being able to attend to the various and multiform duties of life, the invalid is himself, the subject of constant care and anxiety to his friends, who, by excessive watchings and fatigue, are often themselves, brought to a similar condition.

Think, too, of the loss of time and treasure, caused by sickness. By a somewhat careful calculation, I have arrived at the conclusion, that, taking the population of the world, and deducting one-half for infancy, childhood and old age, the loss of time occasioned by sickness, amounts in the aggregate, to not less than five hundred thousand years annually, or nearly fourteen hundred years daily; which, at the usual value of labor, adding the expense of medical attendance and nursing, amounts to more than five hundred million dollars.

Fred.—I am startled, Father, at your statements! And is all the suffering you have described, and this enormous waste of time and treasure, the result of violating the laws of life and health?

Father.—It is, with the exceptions I have already made. And do you think God will hold that person guiltless, who, by violating the physical laws of his being, is contributing to this enormous sum of human wretchedness? I again assure you, my children, that God's laws, whether physical or moral, cannot with impunity be violated or trifled with. And the penalty of violating the one is as sure as that of the other.

But I cannot now dwell longer upon the *physical* effects of violating the laws of health, as I wish to say a few words relative to the effects of physical disease upon mental action. In other words, the effects of disease of the body upon the mind.

It is a well established fact in physiology, that there is a strong and intimate connection and sympathy existing between man's physical and mental nature,—or between the body and mind, and that what affects the former either favorably or disastrously, will produce a like effect upon the latter. A powerful, vigorous, and well-balanced mind, exists only with a well-developed and healthy body. The nerves are the medium through which the mind acts, and when they become debilitated and shattered by disease, their capacity for mental action is as much diminished as for physical. Here is a man, whose muscular and nervous systems have been greatly affected by long continued disease. His muscles are weak and trembling, his nerves excitable and sensitive. Now if he be a business man, can he, in this condition, lay deep and sagacious plans for business operations, or if laid, can he execute them with the energy necessary to a successful result?

Another suffers with constant pain. If he be a man of literary pursuits, can he, while stung with pain, concentrate his mind upon literary subjects, or prosecute his scientific investigations with any degree of success? If a minister of the gospel, can he make due preparation for the duties of the pulpit? or if a lecturer, can he prepare to interest and instruct the lyceum or the popular assembly?

As the mind is dependent on the body as a basis of action, it is not reasonable to suppose that the greatest mental vigor, is compatible with a weak, physical constitution. Hence we find mental imbecility coincident with physical deformity.

In all ages of the world, the greatest and most lasting good has been exerted by those through whose veins coursed the currents of life uncontaminated and unimpeded by disease.

Effects of disease.

Fred.—Do we not sometimes see persons of diseased and deformed physical structures, possessed of vigorous intellects, and who accomplish much in the world?

Father.—We do. But these are exceptions to the rule, and prove nothing against our position. In such cases, we can but regret that such minds should be thus clogged; and reflect upon what they might have accomplished, if connected with a sound healthy body.

Another fact of importance, and the last I shall name on this point is, that millions of insane persons, who crowd Lunatic Asylums, are rendered such by physical disease.

Frank.—How can it be known, Father, that disease is the cause of insanity?

Father.—The proof is, that restoring the health of the body by medical treatment, often restores the mind to its natural condition. Several cases of partial, and one of total insanity, have come under my treatment; the latter—a lady, previous to my seeing her, had thrown herself into a river with the intention of self-destruction, but was discovered and rescued. In all these cases the mind was fully restored by a restoration of the bodily health.

Fred.—From what you have said, it is very evident that disease of the body may affect the mind; but I cannot conceive how the state of the health can affect one's moral or religious condition unfavorably. Indeed, I have heard it said that sickness is favorable to a religious state of mind, and often leads to it. And I recollect, too, of reading a book, a few years since, entitled "The pleasures of sickness."

Father.—I have no doubt that sickness sometimes calls the attention to religious things, and that to a devout person "pleasure" may come from sickness. Still, I think that reflection and observation will make it plain to you, that bodily disease may affect one unfavorably morally, as well as mentally. Have you never known persons naturally cheerful, and good-humored, become fretful and desponding when suffering with certain forms of disease? Such changes I witness almost daily in the practice of my profession. Persons of a devout, charitable, and meek spirit, under the depressing influence of disease that affects the nervous system, often become morose, sensorious and revengeful, unhappy themselves, and rendering all about them so.

A lady of marked mental endowments, in conversation relative to a disease that affects the nervous system with great severity, once made to me the following remark:—"We cannot be very good Christians while suffering with this disease." Although I could not fully endorse the sentiment, yet I was forcibly struck with the remark, knowing as I did, the extreme mental depression produced by the disease.

Examples of the moral effects of disease.

No small proportion of the gloom and melancholy observable in many professed Christians, owes its origin and existence, to a diseased and excitable nervous system. Such are ever writing bitter things against themselves, deploring their sinfulness and vileness, and with prayers and tears, are seeking to conciliate the Divine favor, while at the same time they predict the Divine displeasure. This state of mind, produced by disease, sometimes leads to utter despair of God's goodness and mercy, and to fearful forebodings of the future. That such is the fact I have had abundant proof in my practice as a physician.

Some years since I knew an elderly lady of undoubted piety, and who was noted for her acts of charity, and was greatly beloved by her neighbors. This lady, while in good health, received an injury by being thrown from a carriage, which so affected the nervous system that she became gloomy and desponding, and came to the conclusion that she was the most wicked person living, and that her wickedness had brought, or would bring, utter ruin upon both herself and family, and upon any who might become connected with the family. For weeks and months, she would sit and bemoan her sinfulness which had brought such misery and ruin upon herself, and all connected with her, and was detected in making arrangements to commit suicide. After remaining in this condition for about ten years, her health was restored, when her mind became perfectly sane, and resumed its former cheerful and happy condition, and remained so until the day of her death.

A few months since, a lady came under my treatment, who, though much diseased, was suffering more from distress of mind than pain of body. She was a lady of piety, but had had some slight trouble with a neighbor, which the morbid sensitiveness of her mind —rendered so by disease, had magnified into the "unpardonable sin." The moans and confessions of what she deemed the fatal and unpardonable act, were distressing to hear. She is now restored to nearly her usual health, the recovery of which has entirely removed her despondency, gloom, and dreadful forebodings, and she is now cheerful and happy.

Numerous similar cases might be adduced, all of which go to prove, that bodily disease may, and often does, have a most disastrous effect upon one's moral or religious condition, and utterly disqualifies for the happiness and usefulness which is both the duty and the privilege of every Christian. But I have not time to dwell longer upon this subject.

Fred.—If what you have said is true, Father, it seems that all the suffering endured in this world, both of body and mind, is the result of violated law.

Father.—It is, my son; either in this world or any other. God never designed that man or any other creature he has made, should suffer pain or be unhappy.

Pain a kind admonition.

Although as St. Paul says, "The whole creation groaneth and travaileth in pain together," yet no organ or part of the body of man or beast, has yet been discovered, whose office it is to cause pain. Pain is always a penalty or punishment for wrong doing. Where there is suffering there is sin, whether the suffering be physical or mental. No pain would ever have existed in the universe of God, had there been no infraction of the laws he has established.

You may, therefore, ever regard the first thrill of pain as a kind admonition, assuring you that you are doing, or have done something wrong, which, if continued, endangers health, life and happiness.

Fred.—Then you believe there would have been no pain nor suffering in this world had there been no violation of law.

Father.—I do. Unalloyed pleasure would have been universal.

Fred.—And would there have been no pain in the hour of death?

Father.—We are not sure there would have been any death, had there been no sin. But without entering a theological question, I would remark, that death is not necessarily painful. On the contrary, it is rather a cessation of sensation;—I mean natural death. Death caused by violating the laws of life, will of necessity, be painful.

If what has been stated be true, how enormous is

the guilt, and how fearful the penalty of violating the laws of our being.

Mary.—It certainly is, Father. And I wish we knew more about these laws, so that we should not be so liable to violate them.

Father.—At our next meeting I intend to commence to give you instruction as to the means of obeying the laws of life ;—in other words, the best means of preserving health and avoiding disease.

We must now close this conversation, although I feel that I have but just introduced the various topics upon which we have spoken, to do justice to which would require a volume.

QUESTIONS ON CONVERSATION XXII.

PAGE 328.—What is of interest to thoughtful persons? What is the most important knowledge? What is said of the value of health, and the means of preserving it?

PAGE 329.—What should we believe were we to credit what is often asserted? What is said of such a sentiment? What more is said of this wonderful structure?

PAGE 330.—What facts are stated? What is law? What is statute law? Moral law? Physical law, or law of nature?

PAGE 331.—What are laws of vegetation? What other laws are named? What are those called which particularly belong to our subject? What are the laws of animal life? What more is said of this principle or law? What problem is here solved?

PAGE 332.—What reasons are given? Should we always have health and live to old age, were we never to violate the laws of life? What is it impossible for all to do?

PAGE 333.—What is said of no small portion of mankind? What are disease and death in such cases? What is true, deducting such cases? What fact does the early history of the human race furnish?

PAGE 334.—What now is common for families? What is said of the vast age to which people lived in those times? How do you account for these facts?

PAGE 335.—In what three ways does disease affect us? What is the definition given of health and disease? What may disease do? What is the first, and what is the ultimate effect of disease?

PAGE 336.—What are some of the *physical* effects of disease? What is said of some diseases? What do some forms of disease produce? What is said of rheumatism, hip-disease, spinal disease, palsy, and locked-jaw?

PAGE 337.—For what does disease incapacitate? Give examples. What is said of the loss of time and treasure caused by sickness? To what does it probably amount annually? What daily?

PAGE 338.—What more is said in regard to violating the physical laws of our being? What is a well-established fact in physiology?

PAGE 339.—What is said of the nerves? Give examples. What is said of one who suffers constant pain? What is it not reasonable to suppose? Hence we find what? By whom has the greatest good been exerted?

PAGE 340.—What other fact is named? What is the proof that disease causes insanity?

PAGE 341.—How can disease affect one's moral or religious condition unfavorably?

PAGE 342.—What is said upon the subject on this and the following page? Of what is all suffering the result? What did God never design?

PAGE 344.—What has never been discovered? Of what is pain the penalty? How should pain ever be regarded? What is said of pain in the hour of death? What of the guilt of violating the laws of our being?

CONVERSATION XXIII.

HYGIENE,—THE PRESERVATION OF HEALTH.

NATURE'S MATERIA MEDICA, AIR, LIGHT, ELECTRICITY, TEMPERATURE, EXERCISE, REST, FOOD, SLEEP, BATHING, CLOTHING—SOME OF THESE BRIEFLY CONSIDERED—AIR, AN IMPORTANT AGENT—NECESSITY OF PURE AIR—IMPURITY OF AIR IN LARGE CITIES—IMPORTANCE OF VENTILATING DWELLINGS, CHURCHES, LECTURE ROOMS, ETC.—WHY CITY AIR IS IMPURE—EFFECTS ON HEALTH OF CHANGE OF AIR—EXERCISE, WHY NEEDED—OUT-DOOR EXERCISE—REGULAR OR DAILY EXERCISE REQUIRED—THREE KINDS OF EXERCISE—ITS IMPORTANCE TO FEMALES—CLASSES MOST BENEFITED BY IT—GYMNASTIC EXERCISE—REST, ITS IMPORTANCE.

Father.—At our last conversation we briefly considered the great value of health, and some of the sad effects of disease. We also defined the terms *health* and *disease*, and showed that the former is under the control of law which is inflexible and immutable—to obey which, secures health,—to violate it causes disease. We have now met to consider how we can best make a practical use of these facts,—in other words, how we can secure health and avoid disease; which study or science, as I have previously told you is called HYGIENE.

The laws of health are violated in numerous ways, which I shall point out as we proceed.

To sustain health the body requires Air, Light, Electricity, Temperature, Exercise, Rest, Food, Sleep, Bathing, Clothing, etc. These are Nature's Materia Medica, and I shall speak briefly upon most or all of them, begining with

AIR.

Few agents have so important an effect as air in promoting health or producing disease.

Fred.—How so, Father?

Father.—It is one of the principal means by which animal life is supported, and, as I told you when speaking of the functions of the lungs, it is that without which we can scarcely live a minute.

In respiration a man inhales a gallon of air every minute, which, mingling with the blood, if impure, must contaminate the vital current. Impurities of the air, are also taken up by the absorbents of the skin—the lymphatics, and conveyed to the blood. In this way the blood becomes impure, and disease is the result. When the air is *very* impure, it proves suddenly fatal to life.

It is obvious that many diseases proceed from vitiated air, or an atmosphere charged with deleterious gases. Influenza, cholera, intermittent, remittent, and yellow fevers, so pestilential and fatal, are communicated by an impure air. Villages, towns and cities, have been well-nigh depopulated, as a result of receiving impure air into the circulation through the medium of the lungs. But no air is so immediately fatal to life as that charged with the poisonous carbonic acid gas. And many an one has lost his life by having the air of his sleeping room filled with it by the burning of charcoal as a means of warming the room.

Thus you see the danger arising from poorly ventilated rooms containing crowded assemblies. If the vital portion of a gallon of air be used up each minute by one person, how long can the air remain pure in an unventilated room filled with people, as school and lecture rooms, churches, etc.?

Frank.—Not long, certainly.

Father.—But using up the vital qualities of the air is not the only, nor the principal evil. You know with every breath, a quantity of carbonic acid is exhaled. Thus the air in close rooms not only loses the vital principle—the oxygen, which sustains life, but it is filled with a poisonous gas which destroys life. By this we see the importance to health, that the apartments of dwelling-houses should be well ventilated, especially sleeping rooms. Persons sometimes sleep in small rooms, with doors and windows closed, —and of course, before morning, the vital principle of the air is exhausted, and the poisonous carbon takes its place. Can we wonder that such awake from lethargic slumber with headache, listlessness and languor? Now this is a plain violation of the laws of life and health.

The air of large cities is usually impure. It is rendered so by being often breathed by a dense population, and stagnated by narrow streets and compact houses,—by exhalations from putrid substances, filthy streets, smoke, etc.

Persons coming from the country to large cities, often become seriously affected by the foul, contaminated air. But how much more seriously affected must be the health of those who reside in cities, and constantly breathe impure air. It need be no marvel, that we see, in large cities, so few fresh and healthy, and so many pale and sickly countenances.

A change of air affects the health, in many cases, almost incredibly, by removing from the city to the country. The effect is greater on infants and children than adults,—thousands of whom die annually from the impure air of cities. By removing from the city to the country, many, laboring under the most serious diseases, have quickly recovered, and the whole system frequently undergoes a complete change.

The air of great towns and cities is peculiarly unfavorable to consumptive persons, and those affected with nervous diseases. Pure air is absolutely necessary to the enjoyment of good health, and health becomes poor in proportion as the air we breathe becomes impure.

We live upon air more than upon food and drink. How necessary, therefore, that it should be pure. Upon this subject a sensible writer remarks,—"If fresh air be necessary for those in health, it is still more so for the sick, who often lose their lives for the want of it. No medicine is so beneficial to the sick as fresh air. It is the most reviving of all cordials, if it be administered with prudence. Fresh air is to be let into the chamber gradually, and, if possible, by opening the windows of some other apartment." Much might be said upon the value of pure air, but as I must be brief, I cannot dwell longer upon this subject.

Thus you see, my dear children, the pernicious, and often fatal effects of violating the laws of life in the use of impure air.

Children.—We do, Father; and we believe all you have told us, and will try to profit by it.

EXERCISE.

Father.—Were we to obey all the other laws of our being, we could not enjoy good health, without strict attention to exercise. We are so constituted that it is as much a law of health, and its observance as much a necessity, as the use of food and drink.

There is a physiological fact to which I called your attention when speaking of the circulation of the blood, which is positive proof, that when God made

Reason why exercise is needed.

the human body, he intended it should perform a certain amount of labor, the failure to do which, or to take exercise equivalent, is a plain violation of an important law of health.

Fred.—To what fact do you refer, Father?

Father.—To this:—That although the heart possesses and exerts great power, yet it is evident that it was not intended, and consequently was not endowed with the power necessary, to propel the blood through the body, without the aid afforded by the alternate contraction and relaxation of the muscles acting upon the veins which lie between them, (See page 168). Just about that amount of labor requisite to provide the necessaries of life, is absolutely requisite to a healthful circulation of the blood, and without that labor, it becomes weak and sluggish, and the system cold and torpid.

And this agrees perfectly with the Mosaic history of the creation of Adam, which says, "And the Lord God took the man, and put him into the garden of Eden, *to dress it and to keep it.*" Thus we see that Adam, in his primeval state of innocence, was not to live in idleness.

Exercise strengthens the muscles, expands the chest, gives tone and vigor to the digestive and nutritive organs, increases the appetite, quickens the circulation of the blood, and determines it from the internal organs to the surface, thereby preventing congestion and obstructions,—equalizes the nervous energy, and disposes to quiet and refreshing sleep. It promotes and equalizes animal heat, aids the secretions and excretions, gives strength and firmness to the bones,—in short, it imparts tone and vigor to every part and organ of the body.

By exercise also, the spirits are enlivened, as well as the body strengthened and refreshed. And it is a truth which cannot be controverted, that where exercise is neglected, the energy and strength of the whole system gradually decay, a morbid, irritable state is induced, the powers of the stomach and bowels sustain particular injury, the appetite fails or is vitiated, the digestive fluids are but imperfectly secreted, the muscles become relaxed and debilitated, the nervous system weakened and deranged, and the whole animal economy sinks into disorder.

> "Toil and be strong. By toil the flaccid **nerves**
> Grow firm, and gain a more compacted **tone**;
> The greener juices are by toil subdued,
> Mellow'd, and subtilized; the vapid old
> Expell'd, and all the rancor of the blood."
>
> *Armstrong.*

Much depends on the manner and kind of exercise employed. It should be regular and taken at stated intervals; no day should pass without a reasonable amount of it; but it should not be excessive,—that is, it should not be carried too far, or continued so long as to produce exhaustion. Extremes should be avoided. Persons may injure themselves by too much, as well as too little exercise.

It is also very desirable that exercise should, if possible, be connected with some pleasing occupation, such as gardening, hunting, building, etc.

Active exercise should not be taken immediately after a full meal.

Fred.—Why not, Father?

Father.—To answer your question, I must first state an important physiological fact, which is, that every muscle, organ and part of the body, when engaged in active exercise, or when hard at work, requires, and must have, an extra supply of blood and nervous force, to support and sustain the increased action. And whenever a limb or an organ, increases its exertion, there is an extra amount of both blood and nervous fluid invited or drawn to it. It is therefore evident, that each meal taken into the stomach, requiring vigorous exertion or labor of that organ during the time of its digestion, renders it necessary that an increased amount of blood and nervous energy should also be supplied, which can be done only by a draft from other parts of the system. Now, it is evident that *all* parts cannot be supplied with an extra amount of blood and nervous force at the same time; and as the stomach must have it in order to digest the food, it is clear that vigorous exercise of other parts draws it from the stomach, and thereby interrupts the digestive process.

Three kinds of exercise.

Fred.—I see now why active exercise should not be taken immediately after meals.

Father.—As exercise produces warmth of the body, and often perspiration also, due care should be taken not to cool off too suddenly, by sitting in a draft of air, or for want of adequate clothing.

As to the *time* of taking exercise, the morning is decidedly preferable.

Exercise may be considered as active, passive and mixed. Active exercise is that which requires vigorous exertion of the muscles, as walking, sawing and splitting wood, gardening, sports, etc. Passive exercise is that which requires little or no exertion of the muscles, such as riding in carriages, sailing boats, steam and horse cars, swinging, etc. Mixed exercise is both active and passive, as horse-back riding. The former is adapted to benefit the healthy and robust,—the two latter,—the feeble invalid. Walking, however, is an easy, gentle, and beneficial exercise, which all can indulge in who have the use of their limbs. To receive the greatest benefit, one should walk erect, with the shoulders thrown back, so as to give full play to the lungs. When the attitude is such that the organs are placed in the most natural position, walking is not only beneficial, but becomes a source of pleasure. Active exercise should always be commenced moderately, and never abruptly and with violence.

Frank.—Why so, Father?

Exercise should be commenced moderately.

Father.—For the reason that it is injurious and dangerous, and sometimes proves fatal to life, to *commence* exercise with great violence, and vigorous exertion of the muscles.

Frank.—How can that be?

Father.—It is for the reason, as I have previously explained, (See page 168), that the rapid action of the muscles, pressing upon the veins which lie between them, forces the blood suddenly and rapidly into the heart and lungs, which suddenly crowds the blood-vessels of the lungs with an excess of blood, causing what is called congestion—a state always dangerous, and which may prove suddenly fatal, by producing inflammation, or by the rupture of the engorged blood-vessels of the lungs. If exercise be commenced moderately, the blood-vessels gradually adapt themselves to the increased impetus given to the blood, and no injury or inconvenience results; and after a short time, it may be increased with safety.

In two instances I proved, upon my own person, the truth of what I have stated, by making excessive haste to reach the railroad depot before the cars left; in both of which cases, most violent coughing and distress in the lungs were produced, the unpleasant effects of which continued for weeks;—an act of imprudence I should never dare repeat.

Riding on horse-back is doubtless preferable to any

Best kinds of exercise.

species of exercise not taken on foot, particularly for nervous and dyspeptic persons; and such ought to spend two or three hours daily on horse-back, when the weather is suitable. This exercise tends to produce an equal distribution of the blood through the extremely minute blood-vessels, and prevent an undue accumulation in the internal organs. As riding tends to equalize the circulation, stimulate the skin, and produce a regular action of the bowels, it is admirably adapted as exercise for consumptives, dyspeptics, and nervous invalids. Riding in some kind of an open carriage, is the exercise next preferable for invalids, to riding on horse-back, as a person has the benefit of changing the air and breathing it pure and fresh, which is of great importance. It also gives one a pleasing variety of scenery and of country.

Fred.—Father, what do you think of dancing? is that good exercise?

Father.—I think dancing does more harm than good.

Fred.—What are your reasons for thinking so?

Father.—I do not object to the simple *exercise* of dancing. It is the conditions and accompanying circumstances under which the exercise is taken, that I object to,—such as heated and dusty rooms, air rendered impure by being repeatedly inhaled,—unseasonable hours,—the body in a state of perspiration, which is suddenly checked by going from warm rooms into cold air with thin shoes and dress;—these, with other evils which might be named, show that dancing is better adapted to produce consumption and other fatal diseases, than to promote health and prolong life.

Exercise, of whatever kind, should, if possible, be taken out of door in the open air. This is particularly desirable for females, as they are so confined by domestic duties, that they seldom have the amount of exercise in the open air necessary for the promotion of health.

Females should make it a part of their duty, to exercise and labor as much as possible in the open air. They injure themselves greatly by sitting so much, instead of which, they should, in some way, exercise the whole body, and if in no other way, by walking daily. During the summer season, they should labor two or three hours a day in the flower-garden or kitchen.

The instructions I have given relative to exercise, are applicable principally to students, professional men, persons of sedentary occupations, and females in the higher circles of life. The agriculturist, mechanic, and those engaged in manufacturing pursuits, of course find sufficient exercise in the prosecution of their business.

Fred.—I have been very much interested in what you have told us about exercise. I did not know that health and bodily vigor depended so much upon it.

But, Father, you have not given us any rules by which we can know how much exercise we should take.

Father.—It would be difficult if not impossible to give definite rules applicable to all, as to the amount of exercise each should take, for the reason that the health and constitution of individuals vary so greatly. It should always be enough to produce some degree of fatigue.

The promotion of health by bodily exercise, is receiving great attention from physiologists at the present time; and in various parts of this and other countries, it is reduced to a system, and is taught in establishments, called Gymnasiums. The performance is called gymnastics, and is designed to develop, and give tone and firmness to the muscular system, and indeed to all parts of the body, and also, as before stated, for the preservation and promotion of health. Gymnastics has been introduced into, and is now taught and practised in many schools and colleges.

A gymnasium has recently been established in this city, by Dr. D. Lewis, a gentleman who has devoted much attention to the subject, and is highly successful; and who merits and enjoys the confidence and patronage of the public.

But I cannot dwell longer upon this subject.

Frank.—We cannot tell you, Father, how much we thank you for the instruction you have given us; and to *prove* that we appreciate it, we will show you, by early rising and otherwise, that we intend to reduce it to practice.

Mary.—Yes, Father, and I mean to work in my flower-garden two hours every fair day.

REST.

Father.—We are so constituted that suitable periods of rest are as important a condition of health, as exercise. Our bodies imperatively demand stated intervals of repose. The digestive organs, as I shall hereafter show, after the digestion of each meal, require a season of rest before they are prepared to resume the labor imposed upon them by the succeeding meal. The muscles and brain, after severe taxation, demand quiet and repose as a condition of future effective action. Indeed, man's entire muscular and nervous powers require periodical cessation from the tax levied upon them by daily toil. All classes of the community—persons of all ages, and in every condition of life, require rest. God has clearly indicated this want of our nature, and has most kindly provided for it, by the division of time into alternate day and night, the one for labor, the other for rest. Day and night, labor and rest, are therefore, the order of nature. The labor of the day, whether it be physical or mental, wearies the body and exhausts its energies. Night gives us back the vigor of which the labor of the day has deprived us. How wise and benignant, then, is this division of time,—for amidst the glare and bustle of day how could we sleep,—amidst the darkness of night, how could we labor?

QUESTIONS ON CONVERSATION XXIII.

PAGE 346.—What is hygiene?

PAGE 347.—What does the body require to sustain health? What is said of air? How much air is inhaled each minute? What if it is impure? In what other way are the impurities of the air taken into the blood? What is the result? What fatal diseases are communicated by vitiated air? What has been done as a result?

PAGE 348.—What air is the most immediately fatal to life? By what means have many lost their lives? What is said of poorly ventilated rooms? What of the carbonic acid exhaled? What more is said of oxygen and carbon?

PAGE 349.—What is said of the air of large cities? In what way is it rendered impure? What is said of persons coming from the country? What of those who reside in cities? What is said of a change of air? What more is said of impure air?

PAGE 350.—What is said of the value of fresh air for the sick? How should fresh air be let into the chamber of the sick? What is said of exercise? What is proved by a physiological fact?

PAGE 351.—What is the fact? With what does this agree? What do we thus see? Name some of the benefits of exercise.

PAGE 352.—What are some of the effects of neglect of exercise? What is said of the manner and kind of exercise employed?

PAGE 353.—With what should exercise be connected? When should active exercise not be taken? What physiological fact is stated?

PAGE 354.—What is said of cooling suddenly? How many kinds of exercise are named? Describe each. What is said of walking? How should exercise be commenced?

PAGE 355.—For what reason? Explain why it is dangerous. What if exercise be commenced moderately?

PAGE 356.—What is said of horseback riding? What of riding in an open carriage? What is said of dancing?

PAGE 357.—Where should exercise be taken? For whom is this particularly desirable? Why? What should females do? To whom is this instruction relative to exercise particularly applicable?

PAGE 358.—What is said as to the amount of exercise that should be taken? What are gymnasiums? What is gymnastics, and for what is it designed? Into what has it been introduced?

PAGE 359.—What is said of rest? What is required by the digestive organs, muscles and brain? What do man's entire muscular and nervous powers require? What has God clearly indicated? What is the order of nature? The labor of the day does what? Night does what? What more is said?

CONVERSATION XXIV.

HYGIENE.

Clothing—Age—Climate—Season—Material for clothing—Quantity of clothing—Fashion in dress—Baths, cold bath, sponge bath, warm bath, shower bath—Sleep, its importance—Hours most suitable for sleep—Best means of making sleep refreshing—Material for beds—Best position of the body for sleep.

CLOTHING.

Father.—A due regard to clothing is another important law of health. And to observe that law, three particulars must be attended to—climate, age, and the season of the year.

CLIMATE.

The principal object of clothing, is to protect the body from unfavorable atmospheric influences, such as cold and dampness, and to preserve the temperature of the body at a healthy standard, and no dress, or mode of dress, can be proper, which fails to meet these demands of the system. Persons of course require much more clothing in cold, than in warm climates. Much, however, depends on age, custom, habit and employment.

AGE.

During the period of youth, there is a much more rapid circulation of the blood, and consequently a greater amount of heat generated, than in middle and old age. The former period of life, therefore, requires less clothing than the latter.

SEASON.

Clothing should be adapted to the season of the year—much more being required in winter than in summer. The change should be made gradually, and with great caution. Winter clothing should be put on early in the fall, and worn late in the spring, the quantity being greatest during the winter months.

MATERIAL FOR CLOTHING.

Clothing, in itself, bestows no heat upon the body, it simply prevents the too rapid escape of the heat generated in the system, into the surrounding atmosphere. The materials, therefore, of which clothing is made, are of great importance to health, some being good, others bad conductors of heat. The latter should be selected, particularly for winter clothing.

Fred.—What do you mean, Father, by good and bad conductors of heat. I thought heat would pass through any thing.

Father.—So it will; but it passes more readily through some materials than others.

Fred.—How do you prove that?

Father.—That is very easily demonstrated. Cold is simply the absence of heat. To deprive the body, or any part of it of its natural warmth, produces a sensation of cold. Heat is diffusive and tends to an equilibrium. If two bodies of unequal temperature be placed in contact, the heat passes from the one of higher, to the one of lower temperature, until both are equal.

Fred.—I understand that, Father. But how do you prove that heat passes more rapidly through, or into one substance than another?

Father.—By experiment. If you place your hand on a cold iron or stone, a sensation of cold will be produced instantly; whereas, the sensation will be but slight, if the hand be placed on a garment of the same temperature, made of wool or fur,—for the reason that the former are better conductors of heat than the latter, and more rapidly conduct the heat from the hand, which causes the sensation of cold.

Fred.—I understand it now.

Father.—Well, then, do you not see that clothing made of materials which are good conductors, may carry off the heat from the body too rapidly, while bad conductors confine it to the body, and thereby preserve its warmth? For example: Linen applied to the surface of the body, produces a much greater sensation of cold than woolen, for the reason that it is a better conductor of heat, therefore the former is not so good an article for clothing as the latter.

Fred.—Will you please tell us what are the best materials for clothing?

Father.—*Furs*, and *woolen cloth*, are best adapted, as articles of dress, to persons exposed to great changes of heat and cold. Flannels are particularly beneficial, preventing colds, rheumatism, etc., during the cold part of the year, and also during the warm season, by shielding the system from chills at evening.

Cotton, is smooth in texture, and is well adapted for garments to be worn next the skin, as it produces little friction.

Silk, being a bad conductor of heat, is a good article for clothing.

Linen, being a good conductor, is a poor material for apparel. Even in warm weather, and hot climates, it is unsuitable for dress, as it conducts away the natural warmth of the body too rapidly.

The texture of clothing is of great importance. The material should be of a *porous character*.

Fred.—Why should clothing be porous, Father?

Father.—Do you not recollect, when speaking of the functions of the skin, I stated that insensible perspiration was constantly passing off?

Fred.—I do; but what has that to do with the quality of clothing?

Father.—Very much. If clothing be impervious to perspiration,—that is, if the texture be so close that moisture cannot pass through it, both the inside of the clothing and the surface of the body will soon become damp with retained perspiration, in the same way that the foot becomes wet when wearing an India rubber boot. But this is not all—the impurities, separated from the blood, and thrown off with the insensible perspiration, will be retained in the clothing. Therefore, it should be made of such material as will permit a free passage of the excreted fluids from the surface of the body.

You should also remember what I have repeatedly told you, that oxygen is received through the pores of the skin, and carbonic acid is thrown off by the same medium. Consequently garments made of materials of close texture, as India rubber and some other kinds, intercept their passage and do much harm.

Another particular, of which, perhaps, you have never thought, is worthy of attention. I refer to the fact that the *color* of our clothing is of considerable importance in regulating the temperature of the body.

Frank.—How can color have any effect?

Father.—Dark colored clothing, particularly black, absorbs the sun's rays, and consequently heat, while clothing of light color reflects heat. Dark colored clothing, therefore, is warmest in winter, and light colored is coolest in summer.

QUANTITY OF CLOTHING.

It is impossible to give definite rules as to the precise quantity of clothing each person should wear; as the demand varies with different individuals. Every one should wear the quantity necessary to keep the system warm and comfortable, and each one can best judge for him or herself, what amount is needed. Robust persons require less than the delicate and feeble.

A great error, especially with females, in this changeable climate, is the want of care in adapting the dress to the exigencies of the sudden changes in the weather. This is a subject worthy of attention, especially of invalids. This caution is particularly necessary in the spring of the year, when changes from heat to cold are so frequent and extreme in consequence of the east winds. In this city, the forenoon is often uncomfortably hot, and the afternoon as uncomfortably cold, the change taking place in an hour or two. Now, it is evident, that should persons, especially the feeble and delicate, fail to change the quantity, and if need be the quality of their apparel, as often as the temperature of the weather changes, they must, of necessity, be exposed to chills and severe colds. We need more clothing in the evening, and while asleep, than during the day; for the reason, that in the evening we have less vital energy, and consequently, less heat is generated in the system than in the day,—and exercise of the body and of the brain is suspended during sleep.

FASHION IN DRESS.

With no small portion of the community, especially with females, more importance is attached to fashion in dress than to health and comfort. Persons must dress fashionably, however dangerous it may prove, or ridiculous it may appear! And fashions are changing constantly, regardless of health or comfort. And what vast numbers are yearly carried to an untimely grave by their devotion to fashion!

The most censurable practice in the fashionable world, has been that of wearing the clothes so tight as to interfere with, and obstruct the circulation of the blood and nervous fluid. You can readily perceive, from what I stated when speaking of the circulation, how pernicious, and even fatal to health, must be compression which obstructs the free passage of the blood through the minute vessels.

On the contrary, our dress should be perfectly loose in every part, so as to allow the freest circulation, and also the escape of perspirable matter, and the absorption of the life giving oxygen. Now do you not see that such a practice is a most flagrant violation of the laws of health?

Mary.—I do, Father.

Father.—In regard to *tight* dressing, I am thankful to be able to say, that of late, common sense has, in a measure, triumphed over the tyranny and wickedness of fashion, and the practice has measurably given place to a more rational mode of dress. Thick soled shoes have also taken the place of thin slippers.

BATHS.

Bathing is the application of water of varying degrees of temperature, to the surface of the body. Baths are used as a luxury, for cleanliness, the preservation and promotion of health, and the cure of disease. They are employed cold, warm, or tepid, and of either fresh or salt water.

COLD BATH.

The temperature of the cold bath varies from fifty to seventy degrees, or it should be about the temperature of rivers and ponds in the summer months.

The cold bath is a powerful agent; and should be employed with discretion. It should be used only by persons who possess a good degree of constitutional energy, and to such it is a powerful tonic.

Frank.—What do you mean by tonic?

Father.—A tonic is what promotes solidity and strength of body. A cold bath should be used only when it produces a reaction, and is followed by a warm glow of the surface.

Frank.—What is a reaction, and how can a cold bath produce a warm glow.

Father.—The application of cold water to the surface of the body, causes a sudden contraction of the numerous small blood-vessels of the skin, and the blood is thereby thrown upon the internal organs, which, causing an increased action of the heart, is thrown back to the surface, filling again the blood-vessels, and producing a redness and glow of the skin,—and this is called reaction. And the warm blood returning from the internal parts to the surface, causes the warmth which is called a glow. And this reaction is the criterion by which to decide whether the cold bath is beneficial or injurious. If it occur with a good degree of energy and promptness, it may be regarded as evidence that the person is in a condition to be benefitted by the cold bath. On the contrary, if a reaction does not take place, and a chilliness and languor are felt, and an uncomfortable sensation continues for any length of time, the person may know that to him, the cold bath is injurious and should not be used, at least, in the ordinary way. Such, however, may be able to use to advantage, the *sponge bath* which I shall soon describe.

Fred.—How long should one remain in a cold bath?

Father.—Five minutes is long enough, in ordinary cases, to remain in the cold water; and on coming out, the whole body should be rubbed over briskly with a coarse towel or flesh brush.

SPONGE BATH.

The most simple, and perhaps, on the whole, the most beneficial form of bathing, is the application of water to the surface of the body with a sponge. . The water may be either cold, tepid or warm, to suit the feelings and wants of the bather. This form of bath is particularly adapted to feeble persons, as they can vary the temperature at pleasure, and need uncover but a part of the surface at a time, which, having sponged and wiped with a coarse towel, should be covered warm, and other parts subjected to the same course, until the whole body has thus been passed over. In this way the most delicate can receive the tonic effect of water and friction, with little or no risk. This bath is almost universally applicable, and can be used in winter as well as summer, and none conduces so much to health. The only apparatus needed for this simple mode of bathing, is a washbowl, a sponge and a towel.

It may be practised daily, or every second or third day, or weekly; and immediately after, brisk friction to the whole surface, with a crash towel should be used, until the skin assumes an agreeable glow. This form of bathing is particularly beneficial to delicate persons; and promotes the growth, health, and activity of children.

The best time for this bathing is morning and evening. Those who are subject to nervous affections,—to wakefulness and disturbed sleep, will find this bath, followed by brisk friction of the whole surface, to be of great value in the removal of these symptoms. This form of bathing is of great value in the prevention of colds, sore throat, and similar complaints.

In concluding my remarks upon the cold and sponge baths, I wish to give you a word of caution,—which is; that neither the cold bath, nor the sponge with cold water, should be used when the body is cold or chilly.

THE WARM BATH.

The temperature of a warm bath should be about ninety-five degrees, or about the warmth of the surface of the body. This kind of bath is much better adapted for bathing in general in most seasons of the year, than the cold bath.

The warm bath soothes and tranquilizes the nervous system, warms and expands the small blood-vessels of the surface, by which any undue fulness or congestion of the internal organs is relieved, and the circulation of the blood, and warmth of the body are thereby equalized. It relaxes, softens and purifies the skin, allays irritation, and induces sleep and quiet repose.

This form of bath is particularly beneficial and agreeable to the aged, whose skin has become dry and hard, the circulation poor and sluggish, and the body cold and emaciated. To such, the warm bath, two or three times a week, will afford great comfort, and will be of much service by invigorating the system and retarding the advance of age.

Fred.—How long should one continue in a warm bath?

Father.—In a bath of ninety-five degrees, one may remain fifteen, twenty, twenty-five, or even thirty minutes.

THE SHOWER BATH.

This is an arrangement by which the water falls from the height of a few feet through numerous small holes, upon the head and body. It can be used at any desirable temperature. This form of bath produces a shock to the system of greater or less severity, according to the temperature of the water.

The strong and robust, who are full of vitality, are the only ones who can bear and profit by the shock of a cold shower bath. The feeble and delicate should never attempt its use. If, however, the water be tepid or warm, the most delicate and frail can use it to advantage. A very pleasant mode of terminating the warm bath is, to step from it into a tepid or moderately cold shower bath for a moment. It gives tone to the relaxed skin, closes the pores, and guards against cold.

Principles explained.

Fred.—I have been listening, Father, with much interest, to what you have said upon the subject of bathing, and I have been expecting you would explain the principles on which it produces such beneficial effects. I cannot understand how bathing, which affects only the surface of the body, can be of so much benefit, for it is the internal organs, you know, that are so liable to disease,—the lungs, liver, kidneys, stomach, bowels, etc. I wish you would explain it.

Father.—Do you not recollect what I told you when we were conversing about the skin and its important functions?

Fred.—You told us many things,—to what do you refer in particular?

Father.—I refer to these facts—that through the skin the body has seven million outlets or drains, to carry off the waste, irritating and poisonous matter, which, if retained in the system would inevitably produce disease of the internal organs.

This waste matter is thrown off by perspiration. It consists principally of carbonic acid, phosphate of lime, urea, and animal oil. Now, if by an unhealthy state of the skin, perspiration be obstructed, and these irritants, which amount to two or three pounds per day, be retained in the blood, they will be thrown upon, or will lodge in the lungs, liver, kidneys, bowels, etc., and produce disease of these organs. The most common cause of the worst forms of dysentery, is a sudden check of perspiration, when, for some time it has been profuse; and many cases also of fever and consumption, owe their origin to the same cause. Now, do you not see the importance of keeping the skin in a condition in which it will perform its functions and carry off these impurities? But this is not all. These agents are liable to collect on the surface and be absorbed, and act as a poison upon the system, producing eruptions upon the skin, inflammation, fever, etc. In all such cases the proper use of some form of bath is precisely what is indicated, to remove from the surface this poisonous animal effluvia, and restore the healthy functions of the skin. Again, when the pores are closed or obstructed, it prevents the absorption of oxygen, consequently the system is deprived of that amount of this vital principle, which, in health, is received through the skin.

When, by a suspension of the functions of the skin, the internal organs have become diseased, it is impossible to restore them to a healthy condition, until the functions of the skin are restored. And in the cure of many diseases, especially fevers, this is more essential than all else; and in many cases is the only thing needed. Do you not, therefore, see the appropriateness and value of baths, both in preserving and restoring health?

Fred.—I do, but I should not have seen it, had you not given this clear explanation.

SLEEP.

Sleep! What is it? Who can define it, or explain its secret power? The true emblem of death, yet the restorer of life's exhausted energies. It suspends, and for the time annihilates, every bodily sense and mental capacity, but it does so, that it may, without molestation, touch with new energy, the over-wrought and weakened springs of nature, and, as it were, galvanize them into new life. It is Nature's anæsthetic, with which she etherizes the system, that she may perform her secret operations, unnoticed and unopposed. "Tired nature's sweet restorer—balmy sleep," is as important to the health and comfort of the human family now, as when the immortal Young sang so sweetly of its value.

Alternate periods of waking and sleep are essential to health. Too little sleep exhausts the spirits, debilitates the nervous system, produces headache, anxiety of mind, and moroseness of temper. Too much debilitates the muscles, relaxes the nerves, and produces a state of indolent stupidity which sometimes is not thrown off for the whole day. It is evident, therefore, that some rules, for regulating the time and hours of sleep, are necessary.

The practice of retiring early to rest and rising early in the morning, is favorable to the development of the powers of the system and the preservation of health. Those who lie too long in bed become effeminate and enervated, and soon lose that activity, which conduces to health, and gives value to life.

Six hours sleep every night is sufficient for any adult person during the summer, who is in health, and in winter seven, or, at the most, eight. Those who indulge for nine or ten hours in bed are commonly wakeful or restless during the first part of the night, and, when they ought to rise, sink to rest and slumber on till a late hour in the morning. The strongest constitution will eventually be injured by such a course, for nothing more certainly destroys the constitution than that of sitting up a great part of the night, and lying in bed the most pleasant and most healthful part of the day, as is too much the custom with those who lead a fashionable life, thereby converting night into day and day into night. This plan of proceeding is sure to injure the health of its votaries and shorten the natural period of life, and it will undermine the strongest constitution, even if accompanied with habits of regularity in other respects. Inverting the established order of things, by turning night into day, soon robs the blooming cheek of its roses and lilies, brings on early decay, and destroys the most vigorous frame.

Persons should not sleep in the same apartments they occupy during the day. The bed-chamber should be spacious and dry, and such as can be exposed to the sun, with the windows open during the day, for the admission of pure air, and the dispersion of the poisonous exhalations of the previous night. The beds should be well shaken up every morning, and these, as well as the bed-clothes, should be fully exposed to the fresh air for a sufficient length of time.

It is safe and proper to allow children to take as much sleep as they please;—and persons far advanced in years, especially if feeble, should indulge in as much quiet repose as they choose to take, as in them the enfeebled springs of life are weakened by overaction and want of sufficient sleep.

"The best means of making sleep refreshing is, to take proper exercise through the day; to avoid strong infusions of tea or coffee in the evening; to make a very light supper, at least an hour or two before retiring to rest, where such a meal is indispensably necessary; to go early to bed; to lie down with a mind as serene and cheerful as possible, placing the body in the position which is most congenial to the feelings and habits of the individual; and to rise betimes in the morning—for it has been observed that most of those who have attained a great age, have been early risers. It must, however, be understood that, although early rising and activity are conducive to health, they should, nevertheless, be regulated by the state of the bodily strength, the season of the year, and the habitual exertions of the mind."

We seldom hear the active laborer complaining of restless nights; these complaints usually come from the slothful and indolent.

"The laborer enjoys more real luxury in sound sleep and plain food, than he who fares sumptuously and reposes on downy pillows, where due exercise is wanting."

Perhaps no class of the community are more prone to violate the laws of health in regard to sleep, and thereby induce disease, than students. They not only often over-tax the brain and nervous system, by "consuming the midnight oil" in vigorous appliance to their books, but they deprive themselves of the only power that can sustain the nervous system—*proper hours of sleep*. And many a student, while devoting his best energies to a preparation for usefulness in future life, has, before closing his college course, by violating this law of health, laid the foundation for premature death.

That sleep may be healthy and refreshing, and may answer its intended purpose, much depends on the material used for beds and bed clothes. It should be sufficiently porous to allow a proper circulation of the air, and of a character which will permit the play of the electric currents.

Frank.—Why should beds and bedding be of such a character?

Father.—For the reason that the heat, which is constantly passing from the body, if it cannot escape through the clothing, becomes excessive; also the air, confined in the bed, is soon deprived of its oxygen,

Best Materials for beds.

it being absorbed by the body,—and is filled with carbonic acid, thrown off by the skin. And what a condition is this for rest, and refreshing slumber—the body deprived of oxygen and surrounded with carbonic acid!

Mary.—What materials are the most objectionable, and what the most suitable for beds?

Father.—The most objectionable are feathers to lie upon, and what are called comfortables for covering, as they confine the air in the bed. Feathers are also non-conductors of electricity, and interfere with the electric currents. The best article for a bed is a hair mattress, and the best coverings are sheets, blankets and other woolen articles. Husks, straw, and cork shavings are preferable to feathers.

There is a somewhat recent invention, called "spring bed" which consists of coiled wires arranged in the manner of the seat of a sofa, which, it is said, secures the softness and elasticity of feathers, without any of their objectionable features. I have never used them myself, but they are highly commended by those who have, and I should recommend a trial of them as a substitute for feathers.

The *position* for sleeping is of consequence. The recumbent is preferable, but not indispensable,—I mean for a healthy person. Some sleep on the back, but this is objectionable, as it is liable to produce nightmare, in consequence of the pressure of the organs of the chest upon the aorta and other vessels. The sick can often lie and sleep on the back better than in any other position. For the healthy, sleeping on the right side is the most favorable. There are physiological causes for its being so.

Frank.—What are the causes, Father?

Father.—The most favorable position for sleeping is that in which the blood and other fluids circulate the most freely and with the least obstruction. When lying on the back, as I have said, the organs of the thorax press upon the aorta and thoracic duct, obstructing the passage of the blood and chyle. Lying on the left side obstructs the free flow of the blood through the left lung, by the pressure of the heart upon it,—the heart being situated to the left of the medial line of the chest. This pressure of the heart also interferes with its own free action, and these are a prominent cause of nightmare. But when lying on the right side, no pressure of one part or organ upon another takes place; consequently, the blood flows freely, producing no unpleasant sensation, nor interfering with the vital functions. I ought to add, that a full stomach, particularly a heavy supper, adds greatly to this pressure and obstruction of the free flow of the blood, and is the most common cause of nightmare. And this is the reason why nightmare occurs more frequently the first, than the latter part of the night.

It is not proper, however, that persons should sleep

How to cure wakefulness.

all night in one position,—they should occasionally turn from one side to the other, and most persons do so unconsciously.

Some persons lie awake a part of the night, for the reason, usually, that they spend more time in bed than nature requires for sleep. Such was the case in early life, with that eminent divine—the Rev. John Wesley. Suspecting the cause to be what I have named, he procured an alarm clock, which he set so as to awake him an hour earlier than usual in the morning. He then retired to rest at nine in the evening, and found he lay awake one hour less than usual during the night. He then arranged the bell to awake him *two* hours earlier, and the result was two hours less wakefulness. Thus he continued the experiment, reducing the time spent in bed, until all was occupied in sleep, which was six hours. And having ascertained the amount of sleep needful to health, his uniform practice, for more than half a century, was to retire at ten in the evening and rise at four in the morning,—an example worthy the imitation of all who wish to economise time, fill up life with usefulness, preserve health, and live, as he did, to a good old age. Thus you see, my children, from what I have said, the importance of obeying the law of health in regard to sleep.

QUESTIONS ON CONVERSATION XXIV.

PAGE 361.—What is said of clothing? What three particulars must be attended to? What is the principal object of clothing? What do persons require?

PAGE 362.—What is said of the period of youth? To what should clothing be adapted? What is said of winter clothing? Clothing does what? What is of great importance to health? What should be selected for winter clothing?

PAGE 363.—What is meant by good and bad conductors of heat? Give the proof, or an illustration. What is farther said of good and bad conductors of heat as materials for clothing?

PAGE 364.—Give an example. What are the best materials for clothing? What is said of flannels? What of cotton? What of silk? What of linen? What of the *texture* of clothing? Why should clothing be porous?

PAGE 365.—What if clothing be impervious to perspiration? What else should you also remember? What is said of the *color* of clothing? Explain the difference between dark and light colored clothing.

PAGE 366.—What is said in regard to the quantity of clothing? What is an error in this changeable climate? When is this caution particularly necessary? What is evident?

PAGE 367.—Why do we need more clothing in the evening and while asleep than during the day? What is said of fashion in dress? What is the most censurable practice? What is fatal to health? How should our dress be? What is such a practice?

PAGE 368.—What is bathing? For what are baths used? How are they employed? What is the temperature of the cold bath? What is said of the use of the cold bath? What is a tonic? Only when should a cold bath be used?

PAGE 369.—What is a reaction? How can a cold bath produce a warm glow? This reaction is what? What if the reaction does not take place? How long should one remain in a cold bath? What should be done on leaving it?

PAGE 370.—What is said of the sponge bath? To whom is it particularly adapted? How often may this bath be used? What should be done immediately after?

PAGE 371.—When is the best time for this bath? To whom is it of great value? What caution is given? What should be the temperature of the warm bath? What effect has the warm bath? To whom is it particularly beneficial?

ANATOMY AND PHYSIOLOGY.—CONVERSATION XXIV.

PAGE 372.—How often should it be used? How long should it be continued? What is said of the cold shower-bath? Who are the only ones that can bear this bath? Who should never attempt its use? What is said of terminating the warm bath? What is its effect?

PAGE 373.—What facts are stated? How is this waste matter thrown off? Of what does it consist? To what does it amount per day? What if it be retained in the blood? What is the most common cause of dysentery, fever, and consumption?

PAGE 374.—To what are these agents liable? What is needed in such cases? Of what is the system deprived when the pores are closed? What more is said of a suspension of the functions of the skin? What is essential in the cure of many diseases?

PAGE 375.—What is said of sleep? What is essential to health? What is the effect of too little sleep? What of too much? What is therefore evident? What is favorable to health? What is the effect of lying too long in bed?

PAGE 376.—How many hours' sleep is required by a person in health? What is said of those who indulge in sleep for nine or ten hours? What is said of inverting the established order of things? What is said of the bed-chamber.

PAGE 377.—What is said of beds and bed-clothes? What is said of children and aged persons? What is the best means of making sleep refreshing? What must be understood?

PAGE 378.—What is said of the laborer? What of students? What of materials used for beds and bed-clothes? Why should beds be of such a character?

PAGE 379.—What are objectionable, and what are the best articles for beds? What is said of the "spring bed"? What is said of the position for sleeping? Why should one not sleep on the back? What is most favorable for the healthy?

PAGE 380.—What physiological causes are given why one position for sleeping is more favorable than another? What is said of a full stomach? What of the cause of nightmare? What is not proper?

PAGE 381.—Why do some persons lie awake a part of the night? What is said of the Rev. John Wesley? What of his example?

CONVERSATION XXV.

DIET.

Appetite, its design and use—Children should be taught the laws of digestion and nutrition—Quantity of food—Excessive eating—Indigestible food—Dr. Beaumont's experiments on digestion—Facts obtained, and their practical benefit—Testimony of medical men—Eating too rapidly—Great variety in food objectionable—Times of eating—Stomach requires rest—Hours for meals—Comparative merits of animal and vegetable food—Swine's flesh, pork, prohibited—Animals most suitable for food—The art of cooking, its errors and evils.

Father.—In a previous conversation, I very briefly explained the process of digestion, and hinted at the results of disregarding the laws by which it is controlled. Regarding the proper observance of dietetic rules of vital importance to health, and their violation as the most prolific cause, directly or indirectly, of a vast majority of the ills to which flesh is heir, I shall now dwell at some length upon this subject.

Of all the numerous violations of the laws of health, none are so common, and none, perhaps, so disastrous as those relating to diet. Intemperance in eating and drinking, produces more suffering, both physical and mental, than any other, and, I think I may say in truth, all other causes.

Let us first consider the subject of eating. Our bodies require a certain amount of nutriment to furnish material for repairing the waste of the system, and, in the young, for the growth of the body also. And as I have already told you, ample means are provided in the digestive fluids for the digestion of that amount of food required for the nutriment of the body, *and no more.* We are also endowed with appetite, for the three following purposes :—1st. To inform us when food is needed. 2d. When enough is taken. 3d. To afford pleasure in eating. Now these three objects, I think, would be fully accomplished, had man never abused his appetite :—but the last named object,—pleasure in eating, has, in most persons, become the ruling principle, and has thrown the other two entirely into the shade, and rendered them nugatory.

To inquire when food is needed, and if the quality is suitable, and how much should be taken, are the last things to be thought of. The delicious flavor, and an abundant supply, are the all absorbing points of interest. Hence, many eat without regard to time or quantity, so long as food tastes good. Whether it is needed to nourish the system, never enters their thoughts. Whether it will promote, or prostrate strength and vigor, is an inquiry with which they never trouble themselves. As to digestion and nutrition being under the control of law, they seem not to have the remotest idea. In action they agree with the poet—

> "Live while you live the epicure will say,
> And seize the pleasures of the present day."

"*Pleasures!*" Yes, sensual, brutal pleasures! No, not brutal,—brutes never degrade themselves with gluttony. They eat to the satisfying of the demands of nature, and no more. To man alone, the terms epicure and glutton are applicable. Humiliating fact! Heir of immortality, yet slave to appetites which degrade and ruin both soul and body.

Now there are causes for this strange violation of our nature, and abuse of God's goodness in adding pleasure to a necessity of our being,—that of eating.

Fred.—What are the causes, Father?

Father.—Want of correct knowledge of the laws of digestion, and of the dreadful penalty annexed to their violation, may be regarded as a primary cause. Children should be instructed in dietetic rules, and the laws of digestion and nutrition at an early period of life. It should be a part of their education, as much as the multiplication table and the rules of arithmetic.

Fred.—Can children be taught the laws of digestion, so as to regulate their diet by them?

Laws of digestion as easily taught as other sciences.

Father.—I think so. Why not? Children are taught, at an early age, that fire will burn,—that if they put their fingers into it, it will cause pain, and produce blisters and sores, and that these are the effects of the violation of a law of life. Why then may they not be taught that when they eat unripe fruit, and it causes pain and disease of the stomach and bowels, they have violated a law of health, of which the pain and disease are a result and penalty; or when they eat indigestible food, or too much of what is digestible, and it lies heavy, or becomes acid in the stomach, and causes headache and nervous and feverish excitement, that these are the effect of a violated law;—or when late and heavy suppers cause restless sleep, frightful dreams, nightmare, and a sense of languor and fatigue on waking in the morning, that these are the result of wrong doing?

Now I think it quite as easy, to teach, not only the laws of digestion and nutrition, but all the laws of our physical being, as to teach the science of astronomy, or chemistry, or philosophy, or any of the more simple natural sciences.

Frank.—What you say, Father, looks reasonable, and I doubt not is true;—will you please instruct us in the rules of diet, and teach us in what the violation of, and obedience to its laws consist?

Father.—I will name some of the most essential things, but have not time to tell you all, for it would require a volume. One of the most important rules of diet, and one which is almost universally violated, relates to

THE QUANTITY OF FOOD.

I said this rule is generally violated. That may seem a grave charge, but I believe it true. A vast majority of the community, eat from one-third to one-half more food than the system requires, or the stomach can digest. Consequently, with them there can be no perfect digestion, for, as I have already said, there is no excess of gastric juice supplied to digest an excess of food. Nature never does her work in this way. Consequently a part of the food remains undigested, which, by fermenting becomes acid and rancid, generating a gas which fills the stomach and bowels, causing flatulence, painful eructations, and often cholic. In some cases the contents of the stomach become so exceedingly acrid and irritating, as to excoriate the mucous membrane of the stomach and throat, and cause rapid decay of the teeth. Diluted nitric acid could scarcely prove more injurious. But the irritation produced by the acrid contents of the stomach, is not the only, nor the greatest evil.

You must recollect that out of this acrid mass is derived the nutriment from which is formed, not only the blood, but all the fluids and solids of the body. And who cannot see, that blood formed of such acid

Effect of acrid material in the blood.

and acrid material, must be exceedingly impure, and instead of forming a sound, healthy body, and imparting a healthful and pleasurable stimulus to the system, must prove a source of irritation, causing headache, flushing, eruptions of the skin, and excitement of the whole nervous system, and poisoning all the fluids of the body. Nor is this all. Excessive eating exerts an equally unfavorable effect upon the mind, causing irritability, moroseness, and depression of spirits, which renders one unhappy, and disqualifies for vigorous mental action.

Fred.—I should think the blood, made of such acrid material, would be a mass of impurity.

Father.—So it would, and would cause immediate death, did not the secreting organs perform an extra amount of labor, in separating from the blood, these impurities. But in doing so, they often become diseased themselves, both by over-exertion, and from the irritation caused by the poisonous impurities while passing through them. Hence excessive eaters usually have more or less derangement and disease of the liver, kidneys, stomach, and other secreting organs. Another mode of violating the laws of health, is the use of *indigestible food*. Upon this subject I might, and to do justice, ought to say much, but time will not allow me now to go into all the minute particulars. I shall only name the most important facts.

The comparative digestibility of different kinds of

Experiment on digestion.

food, has been a subject of study and experiment by many able physiologists.

Fred.—You say of study and experiment,—I do not see how it is possible to experiment on digestion, as that takes place within the stomach, and, of course out of sight.

Father.—It is true, that under ordinary circumstances, we cannot actually see what is going on in the stomach. Yet that advantage has been enjoyed, in a few instances, by which much valuable knowledge has been obtained in reference to the phenomena of digestion.

Fred.—Do you mean that persons have sometimes looked into the stomach, and seen the process of digestion going on?

Father.—A few instances of that kind have occurred.

Fred.—How is that possible?

Father.—By artificial openings into the stomach, the most important case of which occurred in the year 1822. The facts are briefly as follows:—A young man, named Alexis St. Martin, eighteen years of age, of robust health and good constitution, was wounded in the stomach by the accidental discharge of a musket, at the distance of one yard from the muzzle of the gun, loaded with duck-shot, which made an opening into the stomach, the size of a man's hand. The young man recovered from the wound, and enjoyed good

Beaumont's experiments on digestion.

health after, but the orifice or opening never closed, and through it, the process of digestion could be clearly seen.

This afforded a rare opportunity for obtaining knowledge upon the subject, and his physician, Dr. Beaumont, a surgeon in the military service of the United States, did not fail to improve it by making many interesting experiments, by which important information was obtained upon the subject of digestion.

Fred.—That was wonderful indeed! Did not the experiments cause pain?

Father.—Not any. Dr. Beaumont began his experiments in May, 1825, and continued them for several months, St. Martin, having fully recovered, and being then in perfect health.

Mary.—It certainly must have been a wonderful sight to look into the living stomach and see digestion going on. Do you know, Father, if he is yet living?

Father.—I do not. I saw him a few years since at an annual meeting in this city, of the Massachusetts Medical Society, but have heard nothing from him since.

Frank.—What discoveries did Dr. Beaumont make by his experiments?

Father.—Many very important ones, but I cannot stop now to tell you all, as I have much to say upon digestion before we close this conversation. I will,

Table on digestion.

however, name a few. He discovered and demonstrated many important facts relative to the nature and power of the gastric juice. He found, also, that the time required to digest different articles of food varies greatly, and is much affected by the manner of cooking.

Frank.—Will you tell us what the difference is?

Father.—I will give you the mean time required to digest the articles in most common use.

TABLE *showing the mean time of digestion of the different articles of diet.*

ARTICLES OF DIET.	MODE OF PREPARATION.	TIME FOR DIGESTION.
		H. M.
Rice,	Boiled,	1 00
Tripe, soused,	Boiled,	1 30
Eggs, whipped,	Raw,	1 30
Trout, Salmon, fresh,	Boiled or fried,	1 30
Apples, sweet and mellow,	Raw,	1 30
Sago,	Boiled,	1 45
Tapioca,	Boiled,	2 00
Milk,	Boiled,	2 00
Liver, beef's, fresh,	Broiled,	2 00
Eggs, fresh,	Raw,	2 00
Codfish, cured, dry,	Boiled,	2 00
Apples, sour and mellow,	Raw,	2 00
Milk,	Raw,	2 15
Eggs, fresh,	Roasted,	2 15
Turkey, domestic,	Boiled,	2 25
Gelatine,	Boiled,	2 30
Turkey, domestic,	Roasted,	2 30
Lamb, fresh,	Broiled,	2 30
Hash, meat and vegetables,	Warmed,	2 30
Beans, pods,	Boiled,	2 30
Cake, sponge,	Baked,	2 30
Parsnips,	Boiled,	2 30
Potatoes, Irish,	Roasted or baked,	2 45
Chicken, full grown,	Fricassee,	2 45
Custard,	Baked,	2 45
Beef, with salt only,	Boiled,	2 45
Apples, sour and hard,	Raw,	2 50

Table on digestion.

ARTICLES OF DIET.	MODE OF PREPARATION.	TIME FOR DIGESTION
		H. M.
Oysters, fresh,	Raw,	2 55
Eggs, fresh,	Soft boiled,	3 00
Beef, fresh, lean, rare,	Roasted,	3 00
Beef steak,	Broiled,	3 00
Pork, recently salted,	Stewed,	3 00
Mutton, fresh,	Broiled or boiled,	3 00
Chicken soup,	Boiled,	3 00
Cake, corn,	Baked,	3 00
Dumplin, apple,	Boiled,	3 00
Oysters, fresh,	Roasted,	3 15
Pork steak,	Broiled,	3 15
Pork, recently salted,	Broiled,	3 15
Mutton, fresh,	Roasted,	3 15
Bread, corn,	Baked,	3 15
Sausage, fresh,	Broiled,	3 20
Oysters, fresh,	Stewed,	3 30
Beef, with mustard, etc.,	Boiled,	3 30
Butter,	Melted,	3 30
Cheese, old, strong,	Raw,	3 30
Soup, mutton,	Boiled,	3 30
Oyster soup,	Boiled,	3 30
Bread, wheaten, fresh,	Baked,	3 30
Turnips, flat,	Boiled,	3 30
Potatoes, Irish,	Boiled,	3 30
Eggs, fresh,	Hard, boiled or fried,	3 30
Green corn and beans,	Boiled,	3 45
Beet,	Boiled,	3 45
Salmon, salted,	Boiled,	4 00
Veal, fresh,	Broiled,	4 00
Fowl, domestic,	Boiled or roasted,	4 00
Ducks, domestic,	Roasted,	4 00
Pork, recently salted,	Fried,	4 15
Soup, marrow bones,	Boiled,	4 15
Pork, recently salted,	Boiled,	4 30
Veal, fresh,	Fried,	4 30
Cabbage, with vinegar,	Boiled,	4 30
Pork, fat and lean,	Roasted,	5 15
Tendon,	Boiled,	5 30

The result of Dr. Beaumont's experiments are very interesting, but they should not be too much relied upon, as they are greatly modified or influenced by circumstances, such as the keenness of the appetite, the amount of exercise taken, the lapse of time since the preceding meal, the state of the health, the perfection of mastication, and above all, the quantity of food taken in proportion to the gastric juice secreted;—for a *large* quantity may remain undigested for several hours, while a smaller amount of the same articles may be entirely digested in one-half or one-third of the time.

From what I have stated relative to Dr. Beaumont's experiments, you see that the length of time required to digest the various articles used for food, varies from one to five and a half hours, *in a healthy stomach*. In a weak and debilitated one, it, of course, takes a much longer time. Now the practical benefit to be derived from this, is the following :—In selecting food, reference should be had to its digestibility; for, let it be remembered, health depends greatly upon the amount of labor imposed upon the stomach,—if too much, disease will sooner or later follow.

Another violation of the laws of health, in eating, is the practice of compounding a large number of articles into one dish, and using a great variety at the same meal, together with condiments, gravies, pickles, preserves, etc.

To observe the proceedings at the dining table of a fashionable hotel or boarding house, is enough to astonish one at the capacity of endurance with which the stomach is endowed. The following will illustrate. A few years since I was dining at a hotel in a New England city, where the following scene attracted

A great variety objectionable.

my attention. A gentleman sitting opposite, commenced his dinner thus:—First, a dish of soup. This disposed of, he next took into a plate a large quantity of lobster, which he cut up fine,—then a head of lettuce, which he also cut and mixed with the lobster. To this he added a liberal portion of mustard, vinegar and sweet oil, all of which were well incorporated with the former.

Fred.—But what was he going to do with such a compound?

Father.—This was his salad, to give zest to his dinner. Next came a plate liberally loaded with meats, potatoes, vegetables, gravy, etc., which, together with the salad, was hastily disposed of, being washed down with a bottle of porter. Then came hot puddings, pies and jellies to which he did ample justice, and finally finished off the meal with fruits, nuts, raisins, etc. I could but say to myself,—*poor stomach!* how long can you endure such barbarous treatment! Now I do not mean to say that all lovers of good food abuse themselves in this way, but it is a specimen of the manner in which thousands and millions sacrifice good health to what they call good living. And can we wonder that the stomach, thus loaded and goaded, at length is exhausted, and that dyspepsia, liver complaint, and the thousand ills to which such abuse gives rise, render life a scene of suffering?

Fred.—I believe what you say, Father, but I wish to ask if physicians generally entertain the same views.

Father.—I think I agree, substantially, with physicians in general, many of whom have written much upon this subject.

Fred.—Please tell us what they say.

Father.—One writer says, "Almost every malady to which the human frame is subject, is connected with the stomach; and I must own, I never see a fashionable physician mysteriously counting the pulse of a plethoric patient, or, with a silver spoon on his tongue, importantly looking down his red inflamed gullet, but I feel a desire to exclaim, 'Why not tell the poor gentleman at once—Sir, you've eaten too much, you've drunk too much, and you've not taken exercise enough!' That these are the main causes of almost every one's illness, there can be no greater proof than that those savage nations which live actively and temperately have only one great disorder—death. The human frame was not created imperfect—it is we ourselves who have made it so; *there exists no donkey in creation so overladen as our stomachs;* and they groan under the weight so cruelly imposed upon them."

Professor Caldwell, of Transylvania University, says, that, "one American consumes as much food as two Highlanders or two Swiss, although the latter are among the stoutest of the race." "Intemperate eating," says he, "is, perhaps, the most universal fault we commit. We are guilty of it, not occasionally, but habitually, and almost uniformly, from the cradle to the grave. It is the bane alike of our infancy and youth, our maturity and age. It is infinitely more common than intemperance in drinking; and the aggregate of the mischief it does is greater. For every reeling drunkard that disgraces our country, it contains one hundred gluttons—persons, I mean, who eat to excess and suffer by the practice." Again he says, "The frightful mess often consists of all sorts of eatable materials that can be collected and crowded together; and its only measure is the endurance of the appetite and the capacity of the stomach."

He adds, "I do not say that such eating always and every where occurs among us, but I do say that it occurs too frequently." Dr. Abercrombie, one of the most eminent physicians of modern times, says, "I believe that every stomach, not actually impaired by organic disease, will perform its functions if it receive reasonable attention; and when we consider the manner in which diet is generally conducted, both in regard to quantity and to the variety of articles of food and drink which are mixed up into one heterogeneous mass, instead of being astonished at the prevalence of indigestion, our wonder must rather be, that in such circumstances, any stomach is capable of digesting at all. In the regulation of diet, much certainly is to be done in dyspeptic cases, by attention to the quality of the articles that are taken; but I am satisfied that *much more depends upon the quantity;* and I am even disposed to say, that the dyspeptic might be almost independent of any attention to the quality of his diet, if he rigidly observed the necessary restrictions in regard to quantity." These quotations are a fair sample of the concurrent testimony of physicians upon this subject, and I might quote enough such to fill a volume, but I must hasten to other particulars.

Another error in diet is that of *eating too rapidly*. The injury resulting from hasty eating is three-fold. 1st. Good digestion depends greatly on the thorough mastication of the food. If it be swallowed in a crude, coarse state, the labor which should have been performed by the teeth, is imposed upon the stomach. Dr. Beaumont found, by his experiments, that digestion is greatly facilitated by minuteness of division, and retarded by the opposite. But it is impossible to thoroughly masticate food, and grind, or break down its more solid parts, when it is eaten hastily. 2d. There should be a thorough admixture of saliva with the food while mastication is going on, it being one of the digestive fluids. But this admixture is impossible unless the food be well ground, and retained in the mouth for a sufficient length of time. Therefore, when food is swallowed hastily, and in a crude state, the benefit to be derived from this important fluid, is

Rules on diet.

lost. 3d. *Hasty* eating is one of the principal causes of *excessive* eating. When we eat moderately, the sensation of hunger, or the appetite, is gradually satiated, and we are thus informed when a sufficient amount of food has been taken. But if it be eaten hastily, the monitions or indications of the appetite, as to quantity, are unheeded, if not unnoticed, and often, in such cases, the capacity of the stomach for holding, seems to be the only limit to the voracious eating. We Americans have acquired a notoriety with other nations not at all enviable, for our greedy manner of eating;—and I think we are justly chargable with being "a nation of gormandizers."

Fred.—You have told us much, Father, about the evils of excessive and hasty eating. Can you give us any definite rules by which we should be governed in these particulars?

Father.—It would, perhaps, be impossible, in the present state of society, to give rules applicable to all. As I have before said, the appetite was given us for this purpose, but it has been so abused that it is not now always to be trusted. These obstacles, though formidable, are not insurmountable, and I shall give you important rules in regard to diet, as we proceed.

Any rule that will cure *excessive* eating, will, I think, be pretty sure to cure *hasty* eating; for, if we limit ourselves to an adequate, but what might appear a small quantity of food, we shall not be likely to

The practice of two or three courses at dinner objectionable.

dispose of it hastily. I will give one rule which I practised myself, with great benefit, some years ago, when in ill health. It is particularly applicable, and would prove highly beneficial, to students, invalids, and all who lead a sedentary life, and would not injure any one. It is this:—When you sit down to your meal, take upon your plate what your *judgment* dictates as a proper quantity, and all you intend to eat at that meal,—then eat that, *and no more.*

Again, the laws of health are violated, as I have already intimated, by the use of *too great variety at a meal.* The common practice is, particulraly at dinners, to have two or three courses,—the first to consist of meats, potatoes, bread, vegetables, gravies, pickles and condiments. And this certainly is variety enough for any one, and is as great as the laws of health will permit; and if the meal were made of this course alone, little harm would result, and the danger of over-eating would be small. But after this course has been finished, which has satiated the appetite, and furnished all the food the system requires for nutriment, then a second course is taken, of puddings, pies, fruits, preserves, etc., so flavored as to whet the appetite anew, and this second course—I might say second meal, is eaten with a zest equal to the first. Now this is all wrong. The system does not require it, the stomach cannot digest it, and the practice will inevitably weaken the digestive powers, impair the health, and shorten life. If the first course is taken, the second should be omitted—if the second is eaten, the first should not be, for *either* will supply the amount of nutriment required, and is all the stomach can digest.

I would here remark, that I wish it distinctly understood, that what I have said relative to the quantity and quality of food, is not equally applicable to all classes. The laborer, who spends ten or twelve hours daily in vigorous bodily exercise, requires a much greater amount of food, and that which contains a greater percentage of nutriment, than the person of sedentary life.

Fred.—Why so, Father?

Father.—For the reason that the waste of the system is in proportion to its activity. In other words, the expenditure of nutriment is in proportion to the amount of bodily exercise taken; and, as with the laborer, the expenditure is greater than with others, the supply must be also. And in this, as in all other cases, when the demand and supply are equal, all is well. Hence the laborer seldom suffers with indigestion, and dietetic rules for him are less needed than for other portions of the community.

TIMES OF EATING.

Healthy digestion and nutrition require that special attention be paid to times of eating. Upon this subject I should be glad to say much, but time will permit me to say but little. I shall speak mostly of the *principles* by which we should be governed in regard to the times of taking meals, and shall say but little in detail, as it is impossible to prescribe definite rules which are equally applicable to all.

The health, habits and employment of individuals, require that the intervals between meals should vary greatly. The laboring man, being subject to greater waste, will require both more frequent and more copious meals than the sedentary and indolent man; and those who eat but little at a time, should eat at shorter intervals than those who eat heavy meals.

No definite or particular hours for eating has been fixed by nature, and these should be decided by mode of life, age, and constitution, the real wants of the system, as indicated by the appetite, being our guide, rather than any definite number of hours.

In regard to intervals between meals, one important fact should be kept in mind, namely, that the stomach is a muscular organ, and, like all such, requires rest. It cannot be constantly employed in the important function of digestion, without fatigue and exhaustion. Intervals of rest are as necessary to the stomach, as to the limbs and other parts of the body. How long could you employ the hands in hard labor, or the limbs in walking, without rest?

Fred.—Not long certainly;—but the limbs rest during the night. Cannot the stomach do the same?

The stomach requires rest.

Father.—It ought to be allowed to do so. It is what nature intended, and what the stomach requires. But from facts already cited, we see that many persons are employed during the day in imposing upon the stomach a burden, to dispose of which, requires its labor during the whole night;—and the sensation of fatigue and exhaustion with which such persons awake in the morning, is proof of the excessive labor it has performed.

Frank.—Will you tell us, Father, how much rest the stomach requires, and the proper time for obtaining it?

Father.—It should be allowed, at least, two hours for rest between meals, and also the greater part of the night. In a healthy stomach, a meal of proper quantity and quality, is digested in about four hours. Add to this two hours for rest, to enable the stomach to recover its tone before another meal is taken, and we have an interval of six hours between meals, which is a very proper period for persons in active life. For the indolent it should be longer,—while for an invalid, on a restricted diet, and for the young who are growing fast, it should be shorter. No one should eat at stated hours, as a matter of course, whether the system require nourishment or not. As a general rule, no one should eat until the appetite indicates the need of food.

HOURS FOR MEALS.

The proper hours for the different meals, depends greatly on the mode of life and constitution of the individual.

One or two hours should elapse after rising in the morning, before breakfast, and two or three after supper, before retiring to rest. If this be correct, it would be proper for the person who rises at five or half-past, to breakfast at seven, dine at one, take supper at six or seven, and retire to bed at nine or ten.

As persons advance in years, the digestive function becomes enfeebled and sluggish. Consequently a longer time is required for digestion, and meals should be taken at greater intervals. For most persons fifty or sixty years of age, not engaged in active life, two meals a day is better than three,—the first, at eight or nine in the morning,—the second, at three or four in the afternoon.

Change of employment, usually renders a change in diet necessary, both in the time of taking meals, and in the quantity and quality of food taken. Boys who leave the farm or the workshop, and girls the kitchen or cotton mill, for boarding-schools, should reduce the quantity of their food about one-half. It should also be of lighter quality, and taken at longer intervals. A failure to observe these rules, will sooner or later induce disease, and injure, if not ruin, the health and constitution.

| Comparative merits of animal and vegetable food. |

Fred.—I am greatly interested, Father, in this subject. What you have said appears so reasonable that I can easily believe it all true. But you have not told us what *kinds* of food are most productive of health. I should think *that* must be an important consideration in diet.

Father.—It certainly is, and I intended to speak upon the subject, if you had not named it. For the last thirty years there has been much said and written, in this country, upon the comparative merits and healthfulness of animal and vegetable food. About the commencement of the time I have named, Dr. Sylvester Graham, an able physiologist, and vigorous writer and lecturer, caused a great excitement upon the subject of vegetable diet. He took the position that no animal food should ever be used,—that vegetables formed the only aliment suitable for man. This sentiment or theory did not, however, originate with Dr. Graham; it has had its advocates from the early ages of the world. But to him is due the credit of introducing the subject more fully to the attention of the American people, than any one who had preceded him.

Many embraced his views,—more in theory than practice,—and some have written and lectured much upon the subject. In justification of their theory, they quote the fact from the Bible, that vegetable food was what God originally designed for man, and that in his primeval state of innocency, nothing else was needed. One writer considers the grant of animal food to Noah, was only fitted to the degraded state of man after the flood. But it might very properly be inquired, if man was not as deeply degraded *before* the flood, while living on *vegetable food*, when "every imagination of the thoughts of his heart was only evil continually," as he was after the flood, or has been at any subsequent period. Another writer, Dr. Cheyne, who lived about one hundred years ago, says, "Animal food was never intended, but only *permitted* as a curse or punishment." If that sentiment be correct, the inquiry might with propriety be made, why some kinds of animal food were granted, and others prohibited? If animal food was designed "as a curse," the worse the quality, the more sure it was to accomplish the object intended.

Again, those who advocate a purely vegetable diet, refer to examples of great longevity and robust constitutions, among those who use such diet. They might also refer to cases, perhaps, equally remarkable, among the Esquimaux, and other northern savage tribes, who live mostly or wholly upon animal food, and that, too, often, which consists principally of the fat of the seal or the blubber of the whale.

They also assert that "taking the life of an inoffensive and harmless animal, and feeding on his flesh, is incompatible with a state of innocence and purity."

| The grant of animal food. |

But is it not a sufficient answer to such an assertion, to say, that the grant of animal food was from God, and that he knew both the wants of man, and what was "compatible with innocence and purity;"—and would a holy and pure being, who requires purity in man, have given for food, what he knew was "incompatible with innocence and purity?"

Fred.—You tell us, Father, what others say upon the subject, but you do not say what your own belief is. Do you think animal food should be used, or should we live on vegetable diet?

Father.—I will tell you frankly, my children, what my sentiments are upon this important subject. I have carefully studied, and duly considered, the arguments and facts, for and against the use of animal food, with a desire to arrive at the truth. But I have not yet become satisfied that it would be for the promotion of health, or for man's best interest in any sense, to *dispense entirely with animal food*. I believe that a moderate quantity of the right kind or sort of it, in connection with the right kinds of vegetable food, is best for most persons in health, and often in sickness, or when recovering from sickness. Mind you, I say *the right sort!* But the great trouble is, man abuses this, like all other of God's gifts. In accordance with the views I have expressed, relative to the use of animal food, are the statements of that distinguished scholar and commentator, Dr. Adam Clarke.

Animal food necessary since the flood.

In his comments on Gen. ix. 3, he says, "There is no positive evidence that *animal* food was ever used *before* the flood. Noah had the first grant of this kind, and it has been continued to all his posterity ever since. It is not likely that this grant would have been now made if some extraordinary alteration had not taken place in the vegetable world, so as to render its productions less nutritive than they were before; and probably such a change in the constitution of man as to render a grosser and higher diet necessary. We may therefore safely infer that the earth was less productive *after* the flood than it was before, and that the human constitution was greatly impaired by the alterations which had taken place through the whole economy of nature. Morbid debility, induced by an often unfriendly state of the atmosphere, with sore and long-continued labor, would necessarily require a higher nutriment than vegetables could supply. That this was the case appears sufficiently clear from the grant of animal food, which, had it not been indispensably necessary, had not been made. That the constitution of man was then much altered appears in the greatly contracted lives of the post diluvians." Such an opinion, from such a man, is entitled to consideration.

God has designated what kinds of animals, fish, and fowl or birds, are good for food;—(See Deut. chap. xiv). But man, instead of confining himself to

The prohibition disregarded.

a reasonable quantity of the kinds God has designated as suitable, has run into the excessive use of almost all sorts of abominable filth, under the name of animal food.

Now to such license, no one has a greater abhorrence than myself. And I think, in civilized communities, the greatest error and abuse, relative to the grant of animal food, is in the use of vast quantities of swine's flesh,—called pork. I fully believe it is not fit for man to eat,—that it renders the blood impure, and causes a vast amount of disease. And God has said by Moses, "Ye shall not eat their flesh, nor touch their dead carcass."

Yet notwithstanding this positive prohibition, and with the assurance that "they are unclean unto you," even Christian communities feed upon "their dead carcass" with as little scruple, and as much zest, as if God had made a special gift of it to man. I believe those kinds of animals which God has designated as proper for food, *are the only ones* man can use with safety and profit, namely, those which "divide the hoof and chew the cud," and those "fish which have fins and scales."

Fred.—But why should animals that chew the cud and divide the hoof, be granted to man for food, and all others prohibited? Is the flesh of such different in quality from the flesh of all other animals, or does God allow the use of some, and prohibit that of others, simply because he has a right to do so?

Why some animals are granted and others prohibited.

Father.—It is true that God is a *sovereign*, and has a right to say what his rational creatures shall, or shall not do. But while he is a sovereign, he is also "our Father," and never acts without a cause, and that cause is always the dictate of wisdom and goodness. We may, therefore, rest assured, that both the grant and the prohibition were made in reference to man's greatest physical, mental and moral good. In all his acts relating to man, God keeps in view the best interest of both soul and body. Although we cannot comprehend it, we know there is a close connection existing between the body and soul, and that great moral changes take place in the mind in consequence of the influence of the body upon it, and this influence depends greatly upon the aliment upon which the body subsists. God knowing what is best for man, and also the quality of the flesh of the different kinds of animals he has made, has wisely and graciously forbidden what would injure both body and mind, and has allowed what is useful to both.

Fred.—It is doubtless so, Father, but I wish to ask why the flesh of animals that chew the cud is more healthful food than that of other animals?

Father.—Two very satisfactory reasons may be given, 1st. Animals which chew the cud, or ruminate, subsist upon grass, grain and vegetables. 2d. Ruminating animals are provided with *two*, *three* or *four* stomachs. The ox, which term includes all animals

Why one kind of flesh affords a pure aliment, and another impure.

of the beef kind,—the cow, heifer and calf, is provided with four, each of which serves a definite and important purpose in the process of digestion, the food, receiving the successive action of each stomach, is thoroughly macerated and digested, affording the greatest possible amount of the purest chyle or nutritive juices. And flesh formed from such nutriment must be pure, digestible, and nutritious, and free from all taint of disease, and consequently suitable food for man.

An eminent scholar and divine, speaking of the different kinds of animals granted to man for food, says, "The flesh of these animals is universally allowed to be the most wholesome and nutritive. They live on the very best of vegetables; and having several stomachs, their food is well concocted, and the chyle formed from it is the most pure because the best elaborated, as it is well refined before it enters the blood."

Not so with the animals *prohibited* as food. The swine, for instance, is one of the most filthy and gluttonous animals on earth. It will live and fatten upon the grossest and most impure food, such as the butcher's offal, carrion, etc., and it wallows in the most disgusting filth.

Now, who can believe that the flesh of such an animal can afford pure and healthful nutriment for man? God has said, "It is unclean unto you,"—that is, it affords a gross and impure aliment, such as is adapted to render the blood impure, and produce scrofulous, scorbutic and cutaneous diseases.

The same reasons exist for the grant of those kinds of "*fish* for food, which have fins and scales," and the prohibition of those which have not. Of all the fish tribe, those with scales are the most nourishing, and afford the purest nutriment, while the flesh of those without scales, consists largely of fatty, glutinous matter, which is very difficult of digestion, and if digested, forms an impure and inferior quality of nutriment.

I should advise the total disuse, by all classes, of pork and lard. I am sure a great improvement in the health and vigor of both body and mind, would follow. I doubt not the use of pork is a fruitful cause of scrofula, with which so large a portion of the people of this country are afflicted.

Frank.—I believe all you have said, Father, about pork, and I will never eat it again. Yet pork is agreeable and pleasant to the taste.

Father.—So perhaps is the flesh of dogs and horses; but if it is, should we make that a justifiable reason for eating it?

Frank.—By no means! But what kinds of animal food are best, and most healthful?

Father.—Beef and mutton are the principal articles, to which may be added veal, lamb, poultry, and some kinds of venison, particularly the deer. But I wish to be *fully understood* upon the subject of animal food.

Animal food should be used moderately.

I do not advocate a free and indiscriminate use of it. On the contrary, I think its use should be moderate in all cases, and in some it should be entirely dispensed with. That too much animal food, even of the right sort, is used, I have not a doubt. It would be better for the community, especially for those of sedentary life, if the quantity was greatly diminished. The latter should use it but once a day, and then with moderation, while the laboring man may use it to profit, twice a day. The quantity used by all classes should be less in summer than winter. It should be cooked in the most simple manner,—either boiled, roasted or broiled, and seasoned with salt only, and eaten with potatoes, bread, rice, etc.

Mary.—Are berries and fruits healthful, Father?

Father.—They are, if eaten as nature formed them, without sugar, cream, etc. The healthful proportions of acid and sweet, which exist in fruits in their natural state, are destroyed by the addition of sugar or any saccharine principle, and then they ferment in the stomach and produce acidity, which causes flatulency, and often disease of the stomach and bowels. The great trouble in the use of most of our food is, *we wish to improve upon nature*, and are not satisfied to take it in the simple, healthy proportions of the elements of which nature has formed it. We compound and mix together, articles totally dissimilar in their nature, until the qualities of each are so blended with those of others, that the individual taste and properties are lost, and a mass is formed, which, although pleasant to the taste, is difficult of digestion and ruinous of health.

The art of cooking, is based mostly upon the principle of compounding numerous articles into one dish, and thereby rendering them indigestible, and consequently innutritious. The present mode of cooking in fashionable hotels, boarding-houses, and not a few families, I regard as an unmitigated evil, so far as health and happiness are concerned. Instead of surprise that so few in such circles are healthy, the wonder is that any are healthy, and that all are not confirmed dyspeptics, or gloomy hypochondriacs. Well would it be for the world, if the simple methods of the ancients, in preparing food by baking unleavened bread on the hearth, and roasting meat by the fire, should take the place of the present art of forming otherwise healthy articles, into health and life-destroying compounds.

The American people, particularly the female portion, are fast becoming a feeble, puny race. They are not what their ancestors were, and are yearly becoming more and more enervated, and will continue to do so, until they are taught to observe, and actually do obey the laws of their physical being,—laws as fixed and immutable as those which hold the earth in its orbit, or sustain the sun in the firmament.

QUESTIONS ON CONVERSATION XXV.

PAGE 382.—What is said of the violations of the laws of health? What of intemperance in eating and drinking?

PAGE 383.—What do our bodies require? For what are ample means provided? For what three purposes are we endowed with appetite? What is said of the last-named purpose? What are the last things to be thought of? What are the absorbing points of interest? Hence many do what? What never enters their thoughts?

PAGE 384.—About what do they never trouble themselves? As to what have they not the remotest idea? In what should children be instructed at an early period of life?

PAGE 385.—Repeat the substance of what is said on this page.

PAGE 386.—What is said in regard to the quantity of food? What is done by a majority of the community? What is the consequence? What is said of the food which remains undigested? What is formed out of this acrid mass? What is said of the blood formed of such acid and acrid material?

PAGE 387.—What is the effect of excessive eating upon the mind? Why does not the blood, formed of such acrid material, cause immediate death? What is the effect upon the secreting organs? What is said of excessive eaters? What of indigestible food? What has been a subject of study and experiment?

PAGE 388.—How is it possible to experiment on digestion? Give a description of the important case named?

PAGE 389.—For what did this afford a rare opportunity? What is said of Dr. Beaumont? What was the result of his experiments?

PAGE 390.—What facts did he discover and demonstrate? What time is required to digest the articles of food in common use?

PAGE 391.—What circumstances modify the results of Dr. Beaumont's experiments?

PAGE 392.—What varies from one to five and a half hours? What practical benefit is to be derived from this? What other violation of the laws of health is named? What is said of fashionable hotels?

PAGE 393.—What illustration is given? What abuse renders life a scene of suffering?

PAGE 394.—What is said by a writer? What by Professor Caldwell?

PAGE 395.—What by Dr. Abercrombie?

PAGE 396.—What other error in diet is named? What is the threefold injury resulting from hasty eating?

PAGE 397.—What have we Americans acquired? What would be impossible? What is said of the appetite?

ANATOMY AND PHYSIOLOGY.—CONVERSATION XXV.

PAGE 398.—What rule is given for eating, and to whom is it particularly applicable? What is said of the use of too great variety at a meal? What is the common practice at dinners? Of what does the first course consist? Of what does the second consist? What is said of this practice?

PAGE 399.—What is not equally applicable to all? What is said of the laborer? What of the waste of the system? What of times of eating?

PAGE 400.—What general principles, in regard to eating, are given? By what should hours for eating be decided? What important fact should be kept in mind?

PAGE 401.—In what are many persons employed during the day? How much rest does the stomach require? What length of time is required to digest a meal? What more is said about the time of taking meals?

PAGE 402.—On what do the proper hours for meals depend? What directions are given? What is said of persons advanced in years? What is rendered necessary by change of employment? What is farther said?

PAGE 403.—Upon what has much been written for the last thirty years? What is said of Dr. Graham? What is done in justification of this theory?

PAGE 404.—What is stated on this page?

PAGE 405.—What would not be for man's best interest?

PAGE 406.—What was the opinion of Dr. A. Clark upon this subject? What has God designated?

PAGE 407.—But what has man done? What is said of pork? What has God said about our eating swine's flesh? What is done by Christian communities? What are the only kinds of animals man can use for food with safety? What kinds of fish? Why should such animals be granted for food and others prohibited?

PAGE 408.—What is said of both the grant and prohibition? What of the close connection between body and soul? God knowing what was best for man, what has he wisely done? Name the reasons given why the flesh of animals that chew the cud is more healthful for food than that of other animals?

PAGE 409.—What is said of the beeve kind? What of the flesh formed of such nutriment? What is said by an eminent scholar and divine? What is said of animals prohibited as food? What is meant by "It is unclean unto you"?

PAGE 410.—What is said of "fish for food"? What of those without scales? The total disuse of what is advised? What kinds of animal food are best and most healthful?

PAGE 411.—What is said of the use of animal food? How often should it be used? How should it be cooked? What is said of berries and fruits? What is the great trouble in the use of food? What do we compound?

PAGE 412.—What is the result? Upon what is the art of cooking based? What is said of the present fashionable mode of cooking? What would be well for the world? What are the American people fast becoming? What more is said of them?

CONVERSATION XXVI.

HYGIENE.

CAUSES OF DISEASE AND DECAY OF THE TEETH CONSIDERED—DECAYED TEETH INJURIOUS TO HEALTH—DIRECTIONS FOR PRESERVING THE TEETH—TEMPERATURE—SUDDEN CHANGES DETRIMENTAL TO HEALTH—OCCUPATION—SOME OCCUPATIONS FAVORABLE, OTHERS UNFAVORABLE TO HEALTH—OCCUPATION SHOULD BE ADAPTED TO THE TASTE, CONSTITUTION, ETC.—OLD AGE—THE INFIRMITIES OF AGE CONSIDERED—TOBACCO AND ARDENT SPIRITS—THEIR EFFECTS CONSIDERED—PHYSIOLOGY AND HYGIENE SHOULD HOLD A PROMINENT PLACE IN SCHOOL INSTRUCTION.

Father.—My children,—there are a few other subjects relative to the laws of health, upon which I wish to give you instruction at the present interview, and this will conclude this series of conversations. At our last meeting I spoke upon digestion and the conditions necessary to its being faithfully performed, one of the most important of which is, the thorough mastication of the food. And as the teeth are the instruments which perform that important office, and as disease and decay of the teeth, which disqualify them for the duty, are so common,—I might say almost universal,

Decayed teeth often the cause of ill health.

I wish to direct your attention for a few moments, to the subject.

It is a serious fact, one worthy the consideration of the public, that much of the disease of the stomach, bowels, lungs, and the general impurities of the blood which are the fruitful cause of a great variety of diseases, have their origin in a diseased and decayed condition of the teeth. I feel the more desirous to give instruction upon this subject, from the fact that neither you nor the community in general, seem to possess very correct and definite knowledge in reference to it.

Fred.—We shall be very glad, Father, to have instruction upon so important a subject. I have noticed that almost every body has more or less decayed teeth, and even very small children, and some times infants lose their teeth. Can you tell us the cause?

Father.—It does certainly seem very strange, that so important organs as the teeth, which doubtless were designed to last during life, should so soon decay; and especially that decay should commence, as you say, in infancy and childhood,—for such is often the case. There must be a cause for this, which I think is not generally understood.

Frank.—Will you please tell us what it is?

Father.—The cause is doubtless somewhat obscure, but I will state a few things which may aid, in some degree, to solve the problem. Much depends

Cause of early decay of teeth.

upon the health of the child or infant, at the formative period of the teeth. One that is sickly during the time the teeth are forming, usually has those which are easily impaired. The decay of the temporary teeth in young children, is owing to a state or condition of the blood, in which the materials that form the bony structure are deficient in quantity or deteriorated in quality. Hence the teeth which are formed when the blood is in this condition, are imperfect in structure, and unable to resist the action of corrosive agents to which they are exposed, as those do, the bony matter of which is received from pure, healthy blood.

But this early decay of the teeth is not always the result of a sickly, diseased state of the child;—it is often hereditary, like many other diseases of which I spoke in a previous conversation.

This view is fully sustained by the most able dentists. Prof. C. A. Harris, in his work on dentistry, says, "The teeth of the child, like other parts of the body, are apt to resemble those of its parents, so that when those of the father or mother are bad or irregularly arranged, a similar imperfection will generally be found to exist in those of the offspring."

"The teeth of the child, therefore, may be said to depend upon the health of the mother and the aliment from which it derives its subsistence. If the mother be healthy and the nourishment that is given the child be of good quality, its teeth will be dense and compact in their texture, generally well formed and well arranged, and, as a consequence, less liable to be acted on by morbid impressions than those of children of unhealthy mothers, or that subsist upon aliment of a bad quality."

M. Mason, a French dentist, remarks, "Does the child derive its life from parents that are unhealthy? The enamel of the milk teeth will be bad; the teeth themselves will be surcharged with a bluish vapor, and in a short time will be corrupted by humid and putrifying caries (ulcers). When the parents are only weakly or delicate, the enamel of the primary (first) teeth will have a bluish appearance, there will be a tendency in them to dry caries, which does not ordinarily make much progress, and seldom causes pain."

"The teeth, while in a pulpy (soft or forming) state, partake of the health of the organism generally. As that is healthy and strong, or unhealthy and weak, so will the elementary principles of which they are composed be deteriorated or of good quality."

Fred.—What remedy is there for this early, hereditary disease and decay of the teeth?

Father.—None whatever. In this, as in other cases I have named, the children suffer for the violation of the laws of health by the parent; and nothing the child can do, or that can be done for it, can avert the penalty. Hence the responsibility resting upon parents; for, not only their own health, but the health

Cause of decay of the permanent teeth.

and well-being of their offspring depend upon their obedience to the laws of health.

But when decay of the teeth is the result of ill health of the child, the remedy consists in the restoration of the health, together with the treatment and care of the teeth which I shall soon prescribe.

Frank.—I understand your explanation of the cause of the decay of the first set of teeth; but what can cause the decay of the *permanent* teeth?

Father.—The want of cleanliness in allowing an accumulation of extraneous matter and corroding incrustations upon the teeth. Also powerful mineral medicines, acting chemically, thus destroying and breaking up their texture,—and acrid filth in the mouth caused by a foul state of the stomach, which becomes concentrated and deposited upon the teeth. These accumulations, and many other corrosive causes, are sure to take place, unless due preventive means be used.

Dr. F. E. Bond, has the following upon this subject. "The enamel and even the bony structure of the teeth are acted upon very readily by many acids, both vegetable and mineral, which combine with the earthy base, lime, and form new compounds with it, breaking up, of course, the integrity of the organ. The enamel is a crystalline mineral substance, and possesses no vital organization, consequently it is quite as liable to be acted upon by chemical agents while in its normal place, as it would be when separated from the body. It is therefore very easy to perceive that this external defence of the tooth may be very easily penetrated, and the ivory of the organ laid open to the action of alimentary matter and fluids of the mouth. It is from this cause that what is called caries results."

Dr. Brewster, of Paris, says, "Constitutional softness of the teeth; the use of medicines during dentition or in after life; the too free use of acids, which, uniting with the lime in the enamel, destroys its strength." And these, he regards, as the principal causes of the decay of teeth.

Fred.—You say, Father, that disease and decay of the teeth cause disease of the stomach and lungs, and an impure state of the blood. How are these effects produced?

Father.—The disease of the stomach called dyspepsia, may be, and often is, produced, by the inability of the teeth, when diseased, to thoroughly masticate the food; and also by taking into the stomach with the food, acrid and putrid matter discharged from carious teeth, and from ulcers in the gums. And the blood is rendered impure by the absorption of this putrid matter, with the chyle, by the lacteals.

Fred.—I perceive that the stomach and blood may become affected in this way. But how can disease of the teeth affect the lungs?

Neuralgia caused by decayed teeth.

Father.—The heat and moisture in the mouth, will always cause the putrefaction of any extraneous matter which is permitted to lodge upon or between the teeth, causing a fetid breath, and imparting an infectious taint to the air as it passes through the mouth to the lungs. In this way the lungs may become affected by disease of the teeth, and consumption may be the result.

Neuralgia often originates in diseased teeth and ulcerated roots. And when neuralgia of the face exists, the teeth should be thoroughly examined, and all roots and parts of decayed teeth should be removed, as they often cause the trouble when no suspicion of the fact exists. But great care should be taken, not to extract, through mistake, sound, healthy teeth, and thereby suffer the loss, without relief of the pain.

DIRECTIONS FOR PRESERVING THE TEETH.

No part of man's organism, perhaps, requires more constant and unremitting care, to keep it in a healthy condition, than the teeth, and, sad to say, no part receives so little, at least from a large portion of the community. A great error exists in regard to the time when care for the teeth should commence. I refer to the sentiment so common, that the temporary teeth, as they are soon to be displaced by the permanent, are of trifling consequence, and that it matters little whether they remain sound and healthy until

Disease of the first set of teeth injure the second.

removed by the second set, or become blackened by disease, and crumble to pieces long before the permanent teeth make their appearance. This error is productive of much mischief, for, by the decay of the temporary teeth, the gum often becomes ulcerated, the living membrane inflamed, giving rise to abcess of the socket or alveolar process, by which portions of the jaw-bone are destroyed, which exfoliate or scale off, and thus the diseased socket of the temporary teeth, causes disease and decay of the permanent. Again, the new teeth, which are situated immediately beneath the root of the temporary, may receive much injury, before they appear in sight, by contact with the latter while in a diseased condition. Dr. Harris remarks, "The decay and premature loss of the temporary teeth constitutes one of the most frequent causes of irregularity in the arrangement of the permanent, and, if for no other reason than the prevention of this, their preservation, if possible should be secured until they are removed by the economy to give place to their successors."

In addition to the causes I have named, of disease of the temporary teeth, much results from inattention to their cleanliness.

"The particles of food and other extraneous matter that lodge between the teeth and interstices, and along the edges of the gums, if permitted to remain, soon undergo a chemical decomposition, and become a

Means of preserving the teeth.

source of irritation to the latter, vitiating the secretions of the mouth, and rendering them prejudicial to the former.

"As the decay of the teeth is dependent upon the presence of vitiated, acrid and corrosive matter, the means of its prevention are obvious. It consists in frequent and thorough cleanliness, which, to be effectual, should be commenced as soon as the first teeth appear."

The best means for cleansing the teeth, is the use of a brush and pure water, with the addition of a tooth pick for youths and adults. The brush should be used on rising in the morning, and after each meal. For the infant or child, the brush should be soft, for the adult it should be of medium stiffness. The brush should be passed up and down, or lengthwise of the teeth, as well as across. The water should be tepid —neither cold nor hot. The tooth pick should be formed from a goose quill, not from metallic substances, and should be used immediately after each meal, that the particles of food lodged between the teeth may be removed before a chemical change takes place, which renders it corrosive.

The use, as I have described, of the tooth pick, the brush and a plenty of water, will usually, if the health be good, prevent all accumulation of tartar and other corrosives, which decompose and destroy the teeth. Too much cannot be done for the cleanliness of the teeth, or too great care be taken to preserve them from decay. But owing to hereditary or other causes, some teeth will decay, notwithstanding every effort be made to preserve them. In such cases recourse should be had to the services of a faithful, honest and skilful dentist; for, *honest faithfulness* is the most important characteristic that should mark the dentist. And the most important part of his practice is the filling and plugging of carious teeth, which, if done skilfully and faithfully, will often preserve them for many years, but if carelessly and unskilfully performed, will do more harm than good.

If through neglect of cleanliness, the teeth have become incrusted with tartar, and the gums tender, sore or ulcerated, the incrustations should be removed by the dentist, in doing which care should be taken not to injure the enamel; after which the tooth pick and brush should be used as I have directed, to prevent a recurrence of the unhealthy accumulations. Persons whose teeth are subject to decay, should have them carefully examined by a dentist two or three times a year, for the purpose of having the attention bestowed upon them, necessary to their preservation.*

TEMPERATURE.

The natural temperature of the body, as I have already informed you, is about ninety-eight degrees.

---

* A work on the teeth, entitled "Facts for the People," by F. D. Thompson, D. D. S., should be in every family.

Changes of temperature should be gradual.

This is the healthy standard, which cannot be essentially changed without detriment,—although the body is capable of enduring great vicissitudes of temperature, if the change be gradual, and not too long continued. Indeed, the changes from summer to winter, and winter to summer, are as essential to the body, to render it healthy and vigorous, as to the soil, to prepare it for vegetation. It is the *sudden* transition from heat to cold, or a too long exposure, that endangers health and life.

The system is capable, as I have said, of enduring a great degree of cold, if gradually brought under its influence, but a sudden application of cold, especially if the body be in a state of perspiration, is fraught with danger. By closing the pores, perspirable matter is retained, which becomes a source of irritation and disease. As a sensible writer remarks,— "There is no change throughout nature more pernicious, either to animal or vegetable bodies, than from extreme heat to intense cold, or from freezing to sudden thawing. Hence it has been observed, that irritating coughs are never so prevalent as when there are sudden alterations of the weather, and when the air, after having been cold, suddenly becomes warm and damp, and after that assumes a considerable degree of coldness again. These transitions occasion a smaller quantity of matter to be thrown off by perspiration, and the lodgment of a greater proportion of fluids upon the internal parts, which become loaded and obstructed; hence catarrhs, diarrhœa, dysentery and many other diseases."

When an ordinary change of temperature is made gradually, such is the constitution of the healthy human frame, that it bears it with impunity; but when it takes place rapidly, danger arises proportioned to the suddenness of the change. Every change of temperature affects our bodies, therefore due precaution should be taken for the prevention of a sudden check of perspiration when the body has been heated. All sudden transitions from one extreme to another, as I have said, should be avoided. No person, during very warm weather, and when the body is greatly heated, should go into an ice-house, cellar, or cold bath, nor sit on cold stones or damp ground. Colds, rheumatism, consumption, and many diseases of a severe and dangerous character have been produced by such imprudence, and even sudden death has been the consequence.

No one should ever sit in a current of air, especially while in a state of perspiration; and the laboring man, whenever he ceases labor, even for a short period, should resume the clothing he had laid aside during active exercise.

Fred.—I never before realized that so much depended upon the right temperature of our bodies, and that sudden changes were so dangerous.

Importance of right choice of occupation.

OCCUPATION.

Father.—When young persons are about to make choice of a profession or occupation for life, it becomes a very serious question what that occupation shall be, for upon that choice depends, in a great measure, their health, happiness, usefulness and prosperity through life. While some employments are productive of health, and consequently of cheerfulness and elasticity of spirits, others exert a baneful influence upon both body and mind.

Fred.—Will you please tell us, Father, what occupations are injurious to, and what productive of health?

Father.—All sedentary employments, particularly those which require persons to continue long in one position, with but little motion of the body, as that of tailors, shoe-makers, etc., are necessarily detrimental to health.

Some occupations are rendered unhealthy by the air being impregnated with deleterious matters, which are taken into the lungs, and either mingle with the blood, or prove a source of irritation to the mucous membrane of the bronchial tubes and air-cells;—such is the employment of stone-cutters, type-founders, printers, chemists, millers, forgers, miners, glass-blowers, painters, gilders, manfacturers of lead, etc. As to the learned professions, as they are called, Law, Medicine and Divinity, there is little to choose, so far as bodily health is concerned;—perhaps that of the physician and the clergy are the most unfavorable to health, on account of the great exposure of the former in the practice of his profession, and of the constant mental labor, and over exertion of the lungs in public speaking, of the latter.

The most healthful occupations are those which require exercise in the open air, such as farming, gardening, building, surveying, engineering, etc. Agricultural pursuits stand pre-eminent, so far as the promotion of health is concerned.

"Cultivating the ground is every way conducive to health. It not only gives exercise to every part of the body, but the very smell of the earth and fresh herbs revives and cheers the spirits, while the perpetual prospect of something coming to maturity, delights and entertains the mind. We are so formed as to be always pleased with somewhat in prospective, however distant or trivial; hence the happiness that most men feel in planting, sowing, building, etc. These seem to have been the chief employment of the more early ages; and, when kings and conquerors cultivated the ground, there is reason to believe that they knew as well wherein true happiness consisted as we do."

Exercise without doors, in some way, is absolutely necessary to health. Those who neglect it, though they may for a while drag out life, cannot be said to

Bad effects of constant and intense study.

enjoy it. Weak and effeminate, they languish for a few years, and then pass to an untimely grave.

Intense and constant study, is, perhaps, more disastrous to health than any other employment. It always implies a sedentary life, and the want of exercise joined to intense thinking, must always be injurious to health.

Close application to study, for only a few months, will sometimes ruin a good constitution, by inducing nervous complaints which can never be removed. Continued thinking will as soon wear out the system as perpetual labor. The whole vital motions may be accelerated or retarded to almost any degree, by the power of the mind,—so great is its influence over the body. Thus, cheerfulness and mirth, quicken the circulation and promote all the secretions; while sadness and profound thought, are sure to retard them.

Thinking, like every thing else, when carried to excess, becomes a vice; and nothing affords a greater proof of wisdom than for a man frequently to unbend his mind by mixing in cheerful company, active diversions, and a change of scenery.

Merchants, clerks, and indeed all classes of citizens who reside in large cities, and are much confined during the day, should rise very early, and walk or ride into the country as often as possible.

Parents should select for their children that occupation or profession to which they are best adapted by their natural temperament, taste, and general constitution of body and mind; and the education of the child should have direct reference to the calling selected. Thousands who die in shops, schools, and colleges, if educated to agriculture or the mechanic arts, would have benefited and blessed the world by long and useful lives.

OLD AGE.

Fred.—I fully agree with all you have said, Father, and I believe health is the greatest earthly blessing we can enjoy, and without it life can hardly be considered a favor. But I wish to ask if you think old age is a blessing? Is it desirable to live to be old?

Father.—What a question, my son! Pray what suggested to your mind such a query?

Fred.—I have heard it remarked, that the infirmities of age are such, and the sufferings consequent upon it so great, that after a man has arrived at seventy years, he is better dead than alive.

Father.—I regard that as a very erroneous and pernicious sentiment. It is true that many in old age suffer from debility and disease. But as a general rule, the infirmities of age, as they are called, are the results of the excesses and irregularities of youth and middle age. They are the deferred execution of a violated law, for, as I have previously told you, it is impossible to violate any of the laws God has estab-

Sufferings of the aged the results of early violated law.

lished, without suffering the penalty,—although the execution of the penalty may be, and often is long postponed. By violating the laws of health in early life, the seeds of disease are deeply sown in the constitution, which, dormant, but not dead, germinate in old age, and bring forth a copious harvest of suffering and sorrow. The violation may long have been forgotten by the offender, but stern justice keeps the record, and ere long the penalty is executed.

This, my children, is the principal cause of the diseases and sufferings of the aged. But these facts are disregarded by the young, and of them it may be said in the language of inspiration, "Because sentence against an evil work is not executed speedily, therefore the heart of the sons of men is fully set in them to do evil."

Fred.—I would like to know what proof there is that the infirmities of age are the effects of wrong doing in early life. It seems strange to me, that the penalty of a violated law should be deferred so long. Why is it not "executed speedily?"

Father.—Because "God is long suffering." "He waits to be gracious" in a physical, as well as moral sense. He has endowed our bodies with the power of enduring the effects of the violation of the laws of health for many years, and if we timely repent and reform, although the health may have been greatly impaired, the recuperative force or vital power, in many

The case of a man one hundred and seven years of age.

cases, repairs the damage and restores the health. But if this violation be persistent, old age pays the penalty, unless premature death cuts short the life of the offender.

I might cite numerous cases in proof of the fact, that men, by violating the laws of health, have brought upon themselves disease and suffering, but when death stared them in the face, they recoiled from his grasp, and reformed their lives, and by ceasing to violate the laws of life, were restored to health, and lived to great age in the possession of vigor of both body and mind. But I will present one case only, which was related by Sir Richard Jebb, late physician to the Royal Family. A particular and very intelligent friend of his, who was about fifty years of age, by excessive living, brought on a train of infirmities which had made great inroads in his constitution, and had induced various disorders, such as colic, gout, spasms, etc., and continued slow fevers; and so wretched was he that he only hoped for death to end his pain and sufferings.

In this condition he consulted Sir Richard, who prescribed for him an exact course of temperance and exercise, by duly attending to which he recovered a sound and perfect state of health.

The following are his own words at the age of one hundred and seven years.

"When," says he. "I resolved firmly to live a temperate life, I soon found myself entirely freed from all my complaints, and have continued so even to this day. I am now convinced that we should consider a regular life as a physician, since it preserves us in health, makes us live sound and hearty to a great age, and prevents us dying of sickness through a corruption of humors. I am now as well as ever I was in my life, (and perhaps better), I even now relish every enjoyment of life better than when I was young. I sleep every night soundly and quietly, and all my dreams are pleasant and agreeable.

"The surest way is to embrace sobriety. What I call a regular and sober life is, not to eat and drink such things as disagree with the stomach, nor to eat or drink more than the stomach can easily digest.

"Some will say that a sober life may indeed keep a man in health, but cannot prolong life. This I know to be false, for I am myself a living instance of it. Oh! what a difference have I found between a regular and an irregular life; one gives health and longevity, the other disease and untimely death. I am now, at the age of one hundred and seven years, hearty and happy, eating with a good appetite and sleeping soundly. My senses are likewise as good as ever they were, my understanding as clear and bright as ever, my judgment is sound, my memory tenacious, my spirits good, and my voice strong and sonorous.

"I likewise enjoy the satisfaction of conversing with

men of bright parts and superior understanding, from whom, even at this advanced age, I learn something.

"What a pleasure and comfort it is that, at my time of life, I should be able, without the least fatigue, to study the most important subjects, nor is it possible that any one should grow tired of such delightful enjoyments, which every one else might enjoy by only leading the life I have led."

Now tell me, my children, do you think this old gentleman was unhappy in consequence of age, and that the the thirty-seven years of his life, (the number exceeding seventy) was void of enjoyment, and of no value?

Fred.—I feel a deep interest, Father, in the aged gentleman you have introduced to us. It must have been delightful to see a man of his age so cheerful and happy, and relishing so fully, all the rational enjoyments of life. This one case is enough to prove that there is no truth in what I have so often heard stated,—that old age is a worthless period of life, devoid of all enjoyment, and filled with sorrow and suffering.

Father.—This man's history is very interesting and instructive. It very forcibly illustrates the truth that I have been teaching you,—that health depends wholly upon obedience to the laws that control it,—that pain and disease are always a penalty for the violation of those laws.

Let us briefly review his history. Sir Richard

Effects of wrong doing—a resolve to reform.

Jebb, the physician who relates it, begins by saying, "This friend of mine was extremely partial to what is called good living;"—that is, eating and drinking to gratify the taste regardless of consequences. The result was, at the age of fifty years, "a train of infirmities had made great inroads in his constitution,"—gout and other diseases racked him with constant pain, and he looked forward to death as the only means of relief. What a picture! The culprit, who disregarded and violated the laws of his being, is now writhing under the infliction of the penalty, and looking to death for relief. Poor man! But his is not a solitary case. Thousands to-day are travelling the same road, and will arrive at the same goal, unless, like him, they reform their lives. We have considered the *dark shades* of this picture,—violated law with its results—disease and suffering. Let us now turn to the bright side,—*obedience* to law, and note *its* results.

He says, "I resolved firmly to live a temperate life." This was a right beginning,—a firm resolve to reform,—just what God requires—to "cease to do evil and learn to do well." The penitent was pardoned. Obedience to God's laws gave relief from pain, and brought health, happiness, and long life. When he ceased to sin, he ceased to suffer.

Such, however, is not always the case. The vital powers may become so exhausted by long continued

Tobacco and ardent spirits.

violation of physical law, that reaction may never take place. In such cases the cessation of wrong doing is of no avail. The reactive forces, once exhausted, can never be restored. Like the soldier, mortally wounded on the field of battle, the violator of physical law lingers for a while, and dies;—but not like the soldier, by the hand of an enemy;—his own hand inflicts the fatal blow,—he dies self-assassinated.

Before closing our present interview, I intended to speak upon several other subjects relating to the laws of health,—especially the use of that poisonous, filthy, breath-polluting, health-destroying, disease-producing, nerve-prostrating and soul-paralyzing article—Tobacco;—and its coadjutor,—Ardent Spirit; but I have time to say but a few words.

The amount of health destroyed, poverty and wretchedness produced, money and time wasted, by the use of these articles, can never be computed in time, and can be known only in eternity.

In the culture and manufacture of tobacco alone, in the United States, in the year 1840, when there was a population of only seventeen millions, there were employed no less than one million five hundred thousand persons, and the consumption in the States at that time amounted to twenty million dollars annually. At the same ratio, at the present time, about three million persons are employed in its culture and manufacture, and the consumption amounts to forty million dollars.

It is estimated that the city of New York alone pays ten thousand dollars a day for cigars. It is also computed, that in America, twenty thousand persons go to the grave every year, from the use of tobacco.

And what good has it ever done in return for such an outlay and waste of life? None whatever! Its effects are evil, only evil, and that continually. And ardent spirit has been, if possible, more prolific of evil than tobacco. The number of paupers produced, of hearts crushed; the amount of health sacrificed, of time lost, the millions of treasure wasted, and the number of untimely graves filled by it, no arithmetician can compute.

Notwithstanding all that has been done in the cause of temperance by the friends of humanity, during the last thirty years, according to carefully compiled statistics, there are annually sacrificed to strong drink in the United States alone, sixty thousand lives!—one hundred thousand men and women are sent to prison for the same cause,—twenty thousand children are sent to the poor-house,—three hundred murders and four hundred suicides are committed,—two hundred thousand children are made orphans to be provided for by public and private charity; and to produce and sustain this enormous amount of crime and misery, there are yearly expended not less than two hundred million dollars.

Value of instruction.

My dear children, as you value health, peace, happiness and prosperity in this life, and the favor of God and the society of the pure in heaven in the life to come, I entreat you, *never*, *no* NEVER! indulge in the use of these accursed drugs. Shun them as you would a mad dog or the raging plague, for while the latter destroy only the body, the former destroy both soul and body.

Other facts upon which I intended to speak I must defer for the present. At some future time I may resume the subject and give you farther instruction; for certainly there is no subject so intimately connected with your well being,—physical, mental and moral, as that upon which we have been conversing.

Fred.—I believe it is so; and I cannot tell you how highly I prize the instruction you have given us. I can but look back with shame to my ignorance upon the subject when you commenced conversing with us.

Frank.—I can say the same.

Mary.—And I can too, Father, for I did not know then how ignorant I was. I now feel that I know more, both of myself and of our adorable Creator.

Father.—And what knowledge can be compared in value to that? To know ourselves and our Maker is the perfection of knowledge. And I am sure, the more you study the subject upon which we have conversed, the more interest you will feel in it, and the more highly you will prize the knowledge obtained.

| Great want of knowledge upon the subject of physiology. |

Fred.—I do not doubt it, Father, for I have felt more and more interested in the subject, from the very first conversation. Why, it seems to me, that I was almost as ignorant as a brute about the wonderful structure of my own body, and the wisdom and goodness of God displayed in its formation. In fact, I had scarcely bestowed a thought upon it, nor had I any desire to do so. I supposed it must be a revolting subject, fit only to be studied by the surgeon and anatomist. But how mistaken I was. No other study has ever been so delightful to me, or interested me so deeply. Why have I never been taught these things in school? Do you suppose all young people are as ignorant upon this subject as we have been?

Father.—I cannot say whether others know more or less than you have known in regard to their physical structure and the laws which control health; but I have reason to believe that there is a lamentable want of knowledge upon the subject, among both young and old.

You ask why you have not been taught these things in school. It is a very natural and very proper inquiry. The answer is, that this subject has never received that attention in schools, nor has it been assigned the prominent position in our system of education in this country, to which it is entitled, and which its importance demands. People have been content to be ignorant of themselves, while making

Children study every thing but their own bodies.

great effort to obtain knowledge upon all other subjects. Children in schools study geography, and are taught to describe the source, size, direction and length of rivers, and at the same time are grossly ignorant of the circulation of the blood through the arteries and veins of their own bodies. They learn the location of cities, but are ignorant of the location of the heart—the citadel of life. They learn the number of inhabitants of a continent, or of the world, but they have no knowledge of the number of pores of the skin, or air-cells of the lungs, or even of the bones and muscles of the body. They study hydrostatics, but remain ignorant of the fluids of their own system.

They study philosophy which makes them acquainted with the forces of nature,—attraction, repulsion, cohesion, etc., but they have little or no knowledge of the force of the heart or the power of the muscles. They learn from the science of chemistry the affinities which exist, and changes which take place in different substances when placed in contact, but they know little or nothing of the neutralizing effects upon each other, of the acids and alkalies of the body.

They study the principles on which the magnetic telegraph conveys intelligence with lightning speed, but they are ignorant of the fact that the nerves of their own bodies are the most perfect telegraph lines ever established.

An improvement in the public mind.

They study mechanics, but do not know that the mechanism of their bodies as far surpasses all else mechanical, as the works of God excel those of man.

They study the properties of light, but not the organs of vision. In short, they study every thing but the structure and functions of their own bodies, and the laws upon which life and health depend. Now this is wrong. The former they should do, but the latter they should not leave undone; for, which is most important, to know the source and size of a river, and in what direction it flows, or in what direction the blood flows through the body, and that its obstruction, mechanical or otherwise, is ruinous to health? To know the location of cities, or the location of the vital organs? To have knowledge of the forces of nature, or the force of the heart? In short, to know every thing without us, but nothing within us, when freedom from pain, health of body and peace of mind, depend upon a knowledge of the structure and functions of the body, and knowledge and observance of the laws which control them?

I am glad, however, to be able to say, that a great change, in reference to this subject, has taken place in public sentiment, within the last few years. Physiology and hygiene are now taught in many schools, and I trust will soon be in all. These branches of knowledge should take as prominent a place in every school, and should be as faithfully taught, with text-books as fully adapted to the purpose, as geography, grammar, arithmetic, or any other science now taught in public or private schools.

Now, my dear children, we have come to the close of this series of conversations. I have presented to you some of the proofs of the goodness and benevolent designs, as well as the extreme perfection of workmanship evinced in the formation of our bodies. I have also endeavored to impress upon you the fact, that the Creator has committed to *our own care and keeping*, these exquisitely wrought pieces of mechanism,—that he has established laws, under the control of which, every fibre and particle of our bodies are placed,—which laws we can obey or violate, as we choose. In conclusion I will only add, that if what I have said relating to the laws of health be true,—if the Creator has established laws, obedience to which secures health and comfort, and the violation of which is the cause, direct or remote, of all the sickness and suffering endured by the human family, then we are forced to the conclusion, that the hundreds of thousands of Physicians who spend their lives in the treatment of disease, are simply making a laborious effort *to counteract the results of violated law*, and as far as possible to repair the damage upon the health and constitution which has thereby been inflicted;—that the hundreds of millions of dollars paid annually to physicians and for medicines, is thrown away, because

The conclusion.

people are either ignorant of the laws of health, or knowing, deliberately and heedlessly violate them in the gratification of appetite, or the indulgence of vicious propensities and passions. People trifle with their health and the laws which control it, and when it is gone, and disease and pain rack their wasted frame, and ghastly death stares them in the face, then, if possible, they would compass heaven and earth to find a physician to cure them, and would give worlds of treasure, did they possess it, to regain the health thus wantonly thrown away.

QUESTIONS ON CONVERSATION XXVI.

Page 414.—What is a serious fact? Who possess but little knowledge in reference to it? What is the cause of the early decay of the teeth?

Page 415.—What more is said upon the subject? What is said by Prof. C. A. Harris?

Page 416.—What by M. Mason? Is there any remedy for this hereditary decay of the teeth? For what do children suffer? What rests upon parents?

Page 417.—What causes the decay of the permanent teeth? What is said by Dr. F. E. Bond?

Page 418.—What by Dr. Brewster? How does disease and decay of the teeth cause disease of the stomach? How does it render the blood impure?

Page 419.—How can it affect the lungs? What is said of neuralgia? What should be done in such cases? What is said in regard to preserving the teeth? What great error exists?

Page 420.—What is said of this error? What may the new teeth receive? What does Dr. Harris remark? What is the result of inattention to cleanliness?

Page 421.—What is the best means for cleansing the teeth? When should the brush be used? How should it be used? What is said of the water? Of what should the toothpick be formed? When should it be used? What is farther said of the use of the toothpick and brush?

Page 422.—What is said of the dentist and his services? What is the most important part of his practice? What more should be done by the dentist? What is the natural temperature of the body?

Page 423.—What is the body capable of enduring? What is said of sudden changes from heat to cold? What do these transitions occasion?

Page 424.—What mo   change of temperature? What is said about sitting in a   What of clothing?

Page 425.—What is said about occupation? What occupations are injurious to health? How are some occupations rendered unhealthy?— Name them. What is said of the learned professions?

Page 426.—What are the most healthful occupations? What is absolutely necessary to health? What is said of those who neglect it?

ANATOMY AND PHYSIOLOGY.—CONVERSATION XXVI.

PAGE 427.—What of intense study? What of the vital motions? What of cheerfulness? What of sadness? When does thinking become a vice? What is proof of wisdom? What should parents do?

PAGE 428.—What sad results often follow the choice of a wrong occupation? What is said of the infirmities of old age?

PAGE 429.—What of violating the laws of health in early life? By whom are these facts disregarded? Why is the penalty of a violated law deferred so long? With what are our bodies endowed?

PAGE 430.—What if this violation be persistent? Of what fact might numerous cases be presented as proof? What case is presented? State the facts. What was prescribed for him? What was the result of a resolve to live a temperate life? What does he say of himself?

PAGE 432.—What is said of this man's history?

·PAGE 433.—To what was this man partial? What was the result? What is said of thousands of others? What is the bright side of this picture? Is such always the case?

PAGE 434.—What more is here said of the violation of physical law? What is said of tobacco and ardent spirit? What of the culture of tobacco in 1840? What of its culture and consumption at the present time?

PAGE 435.—What is said of ardent spirits? What is sacrificed to strong drink? What is the cost of all this?

PAGE 436.—What is said on this page? What is the perfection of knowledge?

PAGE 437.—What more is said upon the subject? Why have not these things been taught in schools? Of what have people been content to be ignorant?

PAGE 438.—What have children studied and learned to describe in schools? And of what were they ignorant? Rehearse what is here stated.

PAGE 439.—Which is most important to be known, what they learn, or what they fail to learn? What has taken place within a few years? What is said of these branches of knowledge?

PAGE 440.—What is said of the ator? What is said in conclusion?

www.ingramcontent.com/pod-product-compliance
Lightning Source LLC
Chambersburg PA
CBHW051233300426
44114CB00011B/716